Health
and
Low-Frequency
Electromagnetic
Fields

Health
and
Low-Frequency
Electromagnetic
Fields

William Ralph Bennett, Jr.

Designed by Sonia L. Scanlon.

Set in Bodoni type by DEKR Corporation, Woburn, Massachusetts. Printed in the United States of America by Edwards Brothers, Inc., Ann Arbor, Michigan.

Library of Congress Cataloging-in-Publication Data

Bennett, William R. (William Ralph), Jr., 1930–

    Health and low-frequency electromagnetic fields /
William R. Bennett.

      p.   cm.

    Includes bibliographical references and index.

    ISBN 0-300-05763-6

    1. ELF electromagnetic fields—Health aspects.

  2. Electromagnetic fields—Health aspects. I. Title.

  RA569.3.B46   1994

616.9′89—dc20   93-40340

               CIP

A catalogue record for this book is available from the British Library.

The paper in this book meets the guidelines for permanence and durability of the Committee on Production Guidelines for Book Longevity of the Council on Library Resources.

10 9 8 7 6 5 4 3 2 1

To the memory of my father,
William Ralph Bennett
(1904–1983),
the finest teacher I have ever known

Contents

Acknowledgments

This book is based on a study that I made for the Committee on Interagency Radiation Research and Policy Coordination (CIRRPC) and that was coordinated by Oak Ridge Associated Universities (ORAU). Most of the material originally appeared as chapter 2 of an ORAU report, "Health Effects of Low-Frequency Electric and Magnetic Fields," completed June 1992. I am indebted to the other members of the original ORAU panel, especially W. E. Gordon, R. J. Reiter, C. Susskind, and D. Trichopolous, for their helpful comments and to Diane Flack of the ORAU staff for editorial help. I am also grateful to R. K. Adair of Yale University, to E. D. Commins and J. D. Jackson of the University of California at Berkeley, and to J. B. Maguire of the University of Pennsylvania for critical readings of the manuscript. J. D. Jackson checked most of the numerical computations. In addition, I would like to thank the following people for helpful discussions on the technical details covered in the book: Franklin Beebe of Cortronix, Joseph Connell and James Gillies of the Metro-North commuter railroad, Gary Goedde of Cooper Power Systems, P. J. Kindlmann of Yale University, and T. B. Lynn of the Dexsil Corporation. E. R. Adair helped locate a number of obscure references.

I am especially grateful to both CIRRPC and ORAU for financial support of this research and for permission to publish my results. The opinions expressed herein do not necessarily reflect those of the two organizations. Finally, I am particularly indebted to Science Editor Jean Thomson Black, manuscript editor Susan Laity, and the members of the Yale University Press staff for their help in preparing the book and to my wife, Frances C. Bennett, for editorial suggestions and everlasting patience.

Introduction

A series of sensational articles in the *New Yorker* written by Paul Brodeur (1989) generated concern over possible health effects from exposure to electromagnetic fields from power-distribution lines of the type found in most urban environments.[1] The first article was based on epidemiological findings by Nancy Wertheimer and Ed Leeper (1979) that an abnormally high percentage of children dying of leukemia in Denver, Colorado, had lived near power lines and by Brodeur's own later discovery of a "cluster" of cancers on a short street containing a power transformer substation in Guilford, Connecticut. Subsequent warnings by Brodeur and others of possible health effects from computer display terminals and television sets has increased public alarm. As a consequence, a number of public and private utilities, as well as several government agencies, have recently sponsored multimillion-dollar studies of the problem. Indeed, a small new growth industry has already resulted, ranging from private companies selling field meters by the carload to groups of people doing epidemiological and biological research. Some experts in these various fields clearly have a vested interest in seeing that further research is subsidized.

One astonishing point about the epidemiological reports is that the electromagnetic fields they refer to are minuscule compared with the natural, unavoidable fields that we encounter in everyday life. The earth's magnetic field is hundreds of times larger than the fields that allegedly cause childhood leukemia in Denver. Common activities like riding a bicycle, driving a car, or flying in an airplane in the presence of the earth's magnetic field result in electric fields inside the body that are much greater than those coupled in from ordinary power-distribution lines. At a more subtle level, thermal electric fields present at the cell level (merely because we are alive and warm) are enormous when compared with the residual fields coupled into the body from these external sources.

Although many reports in the popular press have been written in an emphatically conclusive and deliberately sensational style, the epidemiological studies upon which they were based have not been as convincing. Reports of *possible* causative associations in the original journal articles are described in the popular press as offering clear proof, and they are sometimes accompanied by emotional pleas like "How many more cancers will it take?" (qtd. in Brodeur 1992). To do what, shut down the entire electric power industry? These articles have led to a large number of lawsuits in our increasingly litigious society, as well as several hastily contrived regulations at the state level that cite maximum allowable field strengths to be permitted near power lines — with the threat of future laws requiring warning labels on electrical appliances. At this point, the evidence of any real health effect from stray electromagnetic fields at power-line frequencies is so marginal that such warnings would trivialize the more serious warnings (like the labels on cigarettes) where substantial risk of harm has been clearly demonstrated. Further, the adoption of legal limits on field strengths near power lines that are based essentially on the current status quo (as has been done already in half a dozen states), is bound to be counterproductive. As an example, in Florida (the first state to enact such legislation) the maximum allowable magnetic field at the edge of the right-of-way *increases* with the line voltage. Although this may be a practical necessity for existing lines, the implication from a health point of view seems to be that a higher value of the magnetic field is all right as long as the line voltage is raised. Similarly, the maximum allowable magnetic fields (about 200 milligauss) near the edge of the right-of-way for a 345-kilovolt transmission line in New York is smaller than the magnetic fields encountered during bursts of acceleration inside electrified Amtrak trains. The logical difficulties with such laws are obvious.

There are two basic problems with the epidemiological studies: marginal statistical accuracy and extreme susceptibility to systematic error. Because so much emphasis has been placed on these studies in the popular press, their sources of error should be discussed.

We do not know from any controlled experiment what the actual risks are, and these risks can be estimated only very approximately from the epidemiological data. As with any counting experiment involving random probabilities of occurrence, we can expect that the number of observed cases will approach the average as the number of recorded events increases and that the expected fluctuation ("standard deviation") from that average will increase as the square root of the total number of cases. Consider two identical population groups. Let us say that group A has developed 46 cases of childhood leukemia and group B has developed 54 cases over the same period. Someone has noticed that group B was located near a power line. We need to know whether the difference ($54 - 46 = 8$ cases) has any statistical significance. All we can do is assume that the two numbers are approximately equal to the mean numbers expected from each population group and treat the situation as one that involves the difference count between two (Poisson) probability distributions having different mean values.

Assuming that the two groups are random and independent, the standard deviation in the difference is the square root of the total number of cases, or $\sqrt{(54 + 46)} = 10$ cases.[2] Hence, the difference in the number of cases between the two groups is less than one standard deviation and is not significant. I should note that the observed difference count would be expected to fall within one standard deviation of the mean only about 67 percent of the time; for the remaining 33 percent it would have been larger than ten. Similarly, the mean difference between the two groups would be expected to fall within *two* standard deviations (about 20 cases in the example) 95 percent of the time. In other words, to be 95-percent confident that there was a real disparity in exposure between the two population groups, there would have to be a difference of more than 20 leukemia cases between them. One major problem with such studies is that enormous population groups are required in order to achieve adequate statistical accuracy with a disease as rare as childhood leukemia. The average incidence of childhood leukemia for children under fifteen in Connecticut, for example, amounts to about 3 new cases in 100,000 per year (Flannery et al. 1992, 108). To reach in one year the total number of cases (100) used in the hypothetical example quoted above, the combined populations of the two groups would have to contain about three million children under the age of fifteen. That number is roughly equal to the total population of Connecticut.

The real case studies are much more complicated — especially those involving leukemia, which is bimodal in its age dependence, affects males more than females, and is less common among blacks than whites. In addition, families with young children often move from one area to another. The control group and the exposed group usually involve different total numbers of people so that normalization is required. To describe the results, epidemiologists tabulate a quantity known as the *relative risk*, which is the risk in the exposed individual divided by the risk in the unexposed individual. By definition this quantity goes to one when there is no difference between the two groups and increases when the risk becomes high. To take an example, for cigarette smoking the relative risk of lung cancer is about 9 or 10;[3] the relative risk reported from childhood leukemia studies associated with magnetic fields from power lines ranges from about 0.5 to 2.5. The definition of relative risk, however, results in a skewed distribution, which can be misleading. For example, a relative risk of 0.5 would imply a *reduction* in cancer cases by the same fraction that they would be increased for a relative risk of 2. The relative-risk factors are estimated by assuming that the requisite probabilities are simply equal to the corresponding fractions of cases in the exposed and control population groups that contract the ailment. The effects of statistical fluctuations are included by assuming that the probabilities are normally distributed and by determining the range in relative risk for a 95-percent confidence interval (see, for example, Gordis and Greenhouse [1992]). Many epidemiologists feel that a relative risk as low as 2 is too small to indicate a real causative association and that one really needs a relative risk of about 10 to demonstrate such an association.

Most people have a poor intuitive sense of what constitutes a random sequence of events. If, let us say, they are asked to simulate the results of a coin-tossing experiment, they usually come up with a simple periodic sequence, like heads-tails-heads-tails . . . , and so on. They know in advance that the average number of heads should approach the average number of tails in a large number of tosses, but they want to arrive at that result too quickly. It is also very boring to try to keep track of what happens over an extended sequence that never repeats itself. But if you actually do the experiment based on a table of random numbers, you find that the net number of heads minus tails slowly wanders between positive and negative values and approaches zero only after a very large number of tosses. During this process an astonishing number of heads or tails may occur in a row. It is particularly surprising to see this result the first time, when you realize that the probability of getting either heads or tails at each toss is completely independent of what happened at any time in the past (assuming you have an honest coin). All you can do in advance is predict what the average outcome should be and estimate the expected limits of fluctuation from this average, assuming you do the experiment over and over. But that gives you no information regarding the net number of heads or tails you might expect to see in a row during any particular experiment. All you can say is that if you did the experiment over and over you would probably have fewer than the square root of the number of tosses 67 percent of the time, fewer than twice that 95 percent of the time, and so on. As applied to the cancer problem, it is not unlikely that you might find half a dozen cases (especially of different kinds of cancer) on one short block, considering the enormous number of short streets in the country that do not yield so large a result. If you happen to live on the particular street that has all the cases, it seems unusual. But from a probability point of view alone, it is likely that a street with that large a number of cases would exist out of the millions in the country. The probability argument, of course, cannot prove that there was no physical (or chemical) cause for the cancer in any particular case.

Sources of systematic error are numerous in epidemiological studies. The population groups are often of different ages and different socioeconomic status. Wealthier people can afford better and more frequent medical care, as well as better food and a cleaner environment to live in. It is therefore not surprising to find that they are less likely to contract diseases than people in lower economic classes. The populations sampled usually have different work environments and spend unequal amounts of time in the "exposed" area, and we seldom know what they are exposed to elsewhere. Johnson and Spitz (1990) note that constructional electricians, who seem to have an unusually high risk factor for cancer, are potentially exposed to chemicals like chlorinated diphenyls and naphthalenes, epoxy and phenolic resins, rubber, solder fumes, and synthetic waxes and varnishes, as well as solvents, soldering fluxes, machine oils, cooling agents, and toxic metals. Nonrandomness in the selection of population groups also provides a major potential source of systematic error. Yet some seemingly clever techniques to introduce randomness merely provide additional sources of systematic error; even random-digit dialing selects an economic class that by definition is capable of affording telephones. Similarly, singling out cancer cases from areas close to power substations or high-tension transmission lines tends to select people from lower economic groups; most people who can afford it move away from such eyesores.

Another potential weakness of any epidemiological study occurs through the possibility of "confounding factors" — unknown agents present in the environment that might induce a particular disease and that are correlated with the primary suspected cause. Epidemiological studies point out only correlations, not causes. As seen above, socioeconomic factors can play the role of confounding factors. At the other extreme, chemical carcinogens — preservatives used on telephone poles, oils to cool and insulate power transformers, chemical defoliants — could possibly be associated with the presence of power lines. In extremely high-voltage lines, short-lived corona discharge products like ozone might be found in trace amounts near the line. One author (Kavet 1991) has suggested that electrolysis of metals in water pipes that accidentally carry large ground currents from homes might be involved.

In some instances, cancer may have been induced by insecticides sprayed heavily in areas that were coincidentally used for power substations or transmission lines. The infamous power substation on Meadow Street in Guilford, described in one of Brodeur's *New Yorker* articles ("Calamity on Meadow Street") as a source of stomach cancer, brain cancer, skin cancer, and brain tumors,[4] is in a location previously known to the local inhabitants as Mosquito Alley — a swampy area next to Long Island Sound that has probably been heavily bombarded with insecticide over the years. No studies have reported any correlation between power lines and such pollutants in more than normally occurring levels. Although there seems to be no shortage of *conceivable* confounding factors that might be associated with electric and magnetic fields, cancer caused by such factors should not be blamed on the fields themselves.

In addition to the original Wertheimer-Leeper (1979) study in Denver, major epidemiological studies have been carried out by Savitz et al. (1988), also in the Denver area, and by London et al. (1991) in the Los Angeles area. Although these studies did report statistically significant correlations between increased childhood cancer and so-called high-current wiring geometries, there are inconsistencies between the high-current geometries (that in themselves are not adequately defined) and spot measurements of magnetic fields. In addition, these studies have been criticized by Poole and Trichopoulos (1991) for showing control-selection bias when the cancer exposure was related to residential mobility or socioeconomic status. As we saw, selecting controls on the basis of random-digit dialing (as was done in these studies) tends to exclude poorer people. Positive associations with cancer are usually more prevalent in homes near high-density traffic or in which parental cigarette smoking or abstention from breast-feeding occurs. In cases where the magnetic fields were measured (using a 1-mG cutoff level), relative risks were higher than average in the low-income and low-education groups and lower than average in the high-income and high-education groups.

Two Swedish reports published in 1992 (Feychting and Ahlbom and G. Floderus et al.) suggest a causal association between magnetic fields near high-tension power lines and cancer. These studies were reviewed by J. G. Davis et al. (1993), who note that they supplement the earlier Swedish report by Tomenius (1986) that examines residential exposure in relation to childhood leukemia and brain tumors. Feychting and Ahlbom studied such exposure in relation to brain tumors and leukemia in both children and adults, whereas Floderus et al. concentrated on occupational exposure in relation to these diseases in adults only. The findings in the three studies are mutually contradictory: For childhood leukemia, an inverse correlation was found by Tomenius, in contrast to the positive association reported by Feychting and Ahlbom. For childhood brain tumors, a positive correlation was found by Tomenius, whereas the opposite result was obtained from field data calculated by Feychting and Ahlbom. For adult lymphatic leukemia, the apparent risk increases according to Floderus et al. but decreases in the results of Feychting and Ahlbom. Finally, for adult myeloid leukemia, Feychting and Ahlbom find a positive correlation, in contrast to the opposite result reported by Floderus and colleagues. In addition, self-contradictory conclusions were reached within the study by Feychting and Ahlbom regarding the association between magnetic fields and childhood leukemia and brain tumors, as well as adult myeloid and lymphatic leukemia or adult brain tumors, depending on whether the magnetic fields used were calculated from historical power records or from measured values of the current fields. (The measured values did *not* show a positive correlation.)

One perplexing aspect of most of the reported epidemiological studies is that they have concentrated on population groups exposed to quite low (1 or 2 mG) ambient fields. In contrast, there are population groups (commuters on electrified railroads, for example) who are routinely exposed to average fields that range from about 30 to 130 mG (with peak values of about 600 mG) in whom no unusual occurrence of cancer has thus far been noted. To be fair in the comparison, one should allow for the fraction of time (perhaps 10 percent of the day for many commuters) spent in the train. But the equivalent correction should be made for household exposure, too. Most people do not spend more than half the day in their own homes; in addition, many spend time walking on urban streets where fields of a few milligauss are fairly common. The few studies that were based on occupational exposure have also been limited by statistical uncertainty and systematic error from confounding factors. At one extreme, an early report of an "epidemic" of breast cancer among male telephone linemen turned out to be based on only two observed cases out of fifty-five thousand employees, both of whom were central office workers and not linemen at all (see Matanoski et al. [1989, 1991]). Other studies involving people working at various electrical occupations without exposure to chemical carcinogens have found either negative correlations or small increased relative risk. A major problem with all studies to date is that they have not been based on reliable measurements of the actual exposure to electromagnetic fields or to other carcinogenic agents. It is assumed, for example, that electrical linemen would be exposed to greater fields than almost anyone else; but as with electrical construction workers, linemen spend most of their time working on lines with the voltage turned off.[5]

Finally, there is one basic problem with all these epidemiological studies: they are retrospective rather than prospective. In each case, the information was gathered by checking past histories and often came at secondhand. Its accuracy is subject to the vagaries of personal memory, the recollections of others (the families and friends of the victims, for example), the reliability of public and private records, and similar uncertain sources. Until prospective studies are performed, in which large groups of people are selected at random and their subsequent individual case histories carefully followed, it will be hard to assess the impact of such sources of error on the present studies.

Obviously, epidemiological studies cannot by themselves establish a conclusive cause-and-effect relationship. Indeed, one could argue that because there has *not* been a major, full-blown epidemic of cancer with risk ratios in the region of 10 to 1 that correlated with the increasing consumption of electrical power during this century, the present concern with electromagnetic fields is unjustified. As noted by Jackson (1992), it is remarkable that the number of cases per capita of childhood leukemia and many other forms of cancer has remained relatively constant during a century in which the consumption of electric power per capita has increased exponentially by a factor of more than three hundred. Indeed, if one subtracts the cases of cancer attributed to tobacco from the total number of cancer cases, the number of cases per capita has actually decreased by about 10 percent over the past seventy years. Hence, the alleged cause in the present controversy,

Table Intro. 1. Approximate Variation of Cancer Rates per Capita in Connecticut from 1935 to 1986

Decreased	Roughly Constant	Increased
Liver (f., ÷3)	Uterine	Lung & Trachea (×10)
Stomach (÷3)	Liver (m.)	Skin (×10)
Cervix (÷3)	Mouth & Throat (m.)	Thyroid (m., ×10; f., ×3)
	Rectal	Lymphoma (×5)
	Pancreas (m.)	Brain (m., ×5; f., ×4)
	Ovarian	Testicular (×3)
		Bladder (m., ×3; f., ×1.5)
		Kidney (m., ×3; f., ×2)
		Mouth & Throat (f., ×2)
		Breast (f., ×2)
		Colon (×2)
		Hodgkin's disease (×2)
		Larynx (m., ×2; f., ×4)
		Leukemia (m., ×2; f., ×1.5)
		Prostate (×2)
		Pancreas (f., ×1.3)

Source: Data from Flannery et al. (1992)

Note: All cancers combined increased by about 50 percent. The abbreviation "f." stands for female; "m." for males.

as measured by electrical-power consumption per capita, has gone up by about 30,000 percent over a period when the supposed effect appears to have increased very little. But this argument is oversimplified.

Although the total number of cancer cases per capita has increased by only about 30 percent over the past half-century, some fifteen types of cancer have gone up by factors ranging from about two to ten. Three types (stomach cancer and liver and cervical cancer among females) have decreased by factors of roughly three, and half a dozen other types have remained nearly constant (see table intro.1).

Most cancers preferentially affect older people; one reason for an increase in the total number of cancer cases is the general aging of the population. With most forms of cancer, the incidence rates increase steeply (by factors of a thousand) between the ages of 30 and 60. In Connecticut, the average age (34.2 years in 1980) of the total population (3.3 million) is increasing at the rate of about three years each decade (Flannery et al. 1992). As people live longer, they are more likely to contract cancer of some form. In addition, the increased use of tobacco (including chewing tobacco) and alcohol over the middle portion of this century is taking its toll. Tobacco is believed to be a major risk factor in cancers of the mouth and throat, lung, larynx, cervix, stomach, liver, pancreas, kidneys, bladder, and possibly even leukemia (in veterans) and breast cancer in females. Alcohol is a major risk factor in mouth and throat cancer (especially when used with tobacco) and is associated with cancers of the colon, rectum, liver, breast, larynx, and brain. If one were to ignore all the cases of cancer associated with those two agents alone, there would be few left to blame on electromagnetic fields. Diet (especially involving high fat and in some instances low vitamin A content) is associated with cancers of the colon, rectum, lung, breast, cervix, uterus, prostate, liver, and bladder. Skin cancer is connected with overexposure to sunlight, perhaps early in life. Clearly, there have been enough indulgences in the contemporary American way of life to explain the major increases in cancer shown in table intro.1. Yet the total incidence rate for all forms of cancer (about 400 cases out of 100,000 people per year) is enormous compared with that for childhood leukemia and adult brain cancer, each of which showed fewer than 8 cases in 100,000 per year up to 1986. Hence, because of lower statistical accuracy, it is harder to ascertain causative associations for those two forms of cancer.

Leukemia affects age groups bimodally, primarily attacking older people (where the incidence rate has gone up by a factor of roughly two over the past half-century) and white male children under the age of five (where the peak rate of occurrence is about 8 new cases per 100,000 each year). The rate drops to about 2 per 100,000 from the ages of five to thirty and averages about 3 per 100,000 for the first fifteen years of life. Ionizing radiation (as opposed to extremely low-frequency electromagnetic fields) increases the risk. Down's syndrome is also an established risk factor. Some studies of childhood leukemia show higher incidence rates among higher-income groups. Brain cancer occurs most often in older white males; high-dose X rays, head trauma, and alcohol are associated with increased risk. The rate of occurrence of brain cancer has gone up by a factor of about four for women and five for men over the past fifty years, and there are no known associations with income, diet, or tobacco; reported connections with occupation have been inconsistent.

In some instances, decreases in cancer deaths probably result from the improved efficiency of early detection and treatment enabled by improvements in medical instrumentation, X-ray analysis, computer-automated tomography, and nuclear magnetic-resonance imaging. It should be noted that the magnetic fields used in magnetic-resonance imaging are typically more than ten million times larger than the fields discussed in the studies of power lines.

Although it is well known that direct contact with low-resistance, high-voltage sources of electricity at power-line frequencies can produce severe injury and death by paralysis of the nervous system and intense internal heating, the magnitude of the electric currents coupled into the body when a person walks under a power line is smaller by factors as large as the national debt. Various studies in which live animals were subjected to such fields show no clear evidence of adverse effect: no proof of changes in blood chemistry or the immune system in rats and monkeys; no developmental effects on chick embryos; no deleterious embryonic development in rats and swine. Applications of 60-Hz magnetic fields have shown changes in the circadian rhythms of some animals, rhythms that are probably controlled by secretions like melatonin, which comes from the pineal gland. Yet early reports that melatonin levels in rats were strongly inhibited by static earth-level fields (for example, 1,000 mG) were contradicted by later experiments with very much larger fields. Some data where magnetic fields of 1 G were turned on and off about once per minute showed a statistically significant effect associated with the transient changes. Still, the notion that changes in melatonin levels might produce cancer is mere speculation at present. The fact that a biological effect exists does *not* mean that it is cancerous.

The live-animal studies are at least as prone to statistical uncertainty and systematic error as the epidemiological studies. Of necessity, the number of animals available is small, and the fractional statistical fluctuations are large. In addition to physiological variations among different animals, the behavior of the animals themselves can actually alter the applied fields. Consider an experiment in which rats were placed on the bottom plate of two parallel horizontal conducting plates across which a large voltage was connected to produce an electric field. Because a rat is conductive in comparison with the air between the plates, its very presence will increase the electric field it experiences over its body. This distortion is fairly small when the rat is lying down, but the increase becomes much larger (perhaps by a factor of fifty at head level) if it stands up on its hind legs. When ten or twenty rats are jammed into the same space, the variations are apt to become even more erratic. This effect may explain the inability to obtain reproducible results in some experiments. Studies have been attempted in which the effect of large electric fields on animal behavior and learning patterns is investigated. Although positive results have been reported, they are inconsistent. It is likely that the animals were merely distracted by the presence of the electric fields, which they detect through hair follicle movement, just as humans do. (The threshold for this detection process in humans is quite high, typically about a million volts per meter.) In sum, live-animal experiments have been inconclusive in showing any harmful effects from stray electromagnetic fields at power-line frequencies. Further, most of the experiments reported have involved substantially larger electric or magnetic fields than are experienced near power-distribution lines or common household appliances.[6]

Research at the cellular level has been similarly inconclusive.[7] Experiments with calcium-ion efflux in cells from the brains of chick embryos have produced a variety of results over the 1980s that tend to indicate little more than how difficult it is to find consistent answers. Marginally significant data have been interpreted in terms of different frequency and intensity "windows" (that is, phenomena observed over a limited range of frequencies and not below or above that range), sometimes requiring the simultaneous presence of both electric and magnetic fields. Nearly as many different mechanisms have been proposed to explain the results as there have been papers published on the subject — mechanisms that range from the unlikely (Bose-Einstein condensation) to the totally implausible (cyclotron resonance in viscous liquids: see chapter 6, below). Ironically, the one case where a clear, though *beneficial*, effect has been observed clinically — bone fractures being healed by the application of large, pulsed magnetic fields at low-frequency repetition rates — does not appear to have been studied at the cell level.

Misapprehension has arisen among both the general public and professional biologists from an inadequate understanding of the physical nature of extremely low-frequency electromagnetic fields and of how they might affect the human body. Most people are aware of the dangers of ionizing radiation and of high-power microwave fields, and there has been an unfortunate tendency to assume that similar dangers may occur at power-line frequencies. As a result, some alarmists stress the "dangers" of living in a home where the ambient low-frequency magnetic fields are on the order of a milligauss — levels that are difficult to measure accurately in the physics lab and that are five hundred times smaller than the earth's static magnetic field. Others have suggested that there may be very narrow resonances that would magnify the biological sensitivity to periodic fields enormously and that magically occur at exactly 60 Hz — the standard power-line frequency in the United States. Yet such suggestions have generally been made without discussion of any physically plausible mechanism to provide the required resonances — and without regard for the fact that many other countries use different power-line frequencies (for example, 50 Hz).

Much of the speculation on biological interactions with extremely low-frequency fields has tacitly ignored the fundamental physical laws involved. The physical properties of electromagnetic fields were well described by the work of Maxwell and his predecessors in the nineteenth century, and the solution of Maxwell's equations for most cases of interest is straightforward. Although biological mechanisms can be complicated, they cannot violate the fundamental laws of physics.

My objectives in this book are to describe the general properties of low-frequency electromagnetic fields and the more useful methods for their calculation and measurement and to apply these methods to the determination of fields for representative worst-case limits in the present environment. In addition, I review how these fields are distorted by the presence of the human body and how they couple into it, and I compare the resultant magnitudes with those of unavoidable background fields that arise from natural processes. The relation of these results to experiments dealing with the observation of biological effects from low-frequency fields leads me to conclude that the fields produced inside the body at the cell level from ordinary urban power-distribution lines, household appliances, and computer display terminals are orders of magnitude smaller than the unavoidable background fields produced by natural processes. Insofar as the induced fields are negligible in comparison with such natural sources as thermal noise, they cannot be regarded as a serious health hazard. Although my primary objective is to present and explain these considerations in representative cases, I include enough of the mathematical background for the interested reader to apply the results to different problems and geometries. A variety of physical and biological data has also been included in tabular form throughout the book in order to make the discussion as self-contained as possible. The material thus represents a tutorial on low-frequency electromagnetic phenomena.

Although I believe that the dangers of extremely low-frequency electromagnetic fields to human health have been exaggerated beyond reason, my conclusion is largely based on arguments of plausibility derived from well-established physical principles, as well as from a lack of consistent epidemiological and biological data to the contrary. I by no means contend that further research should not be conducted on electromagnetic interaction with biological processes. Quite the opposite: a number of specific problems that ought to be resolved are discussed in detail. But there does not appear to be any need for a crash research program to investigate low-frequency electromagnetic fields as a cause of cancer. There is no evidence in the data that suggests a danger to human health that compares with the risk produced by cigarette smoking, and there are certainly far more urgent areas for biological research in the public interest. (To name only one, AIDS will probably have killed more than fourteen million people by the turn of the century.)[8] Curtailing electrical facilities through foolishly enacted legislation or a plethora of lawsuits against the power companies and electrical-equipment manufacturers could have a devastating effect on our society. Imagine returning to kerosene lamps or doing away with medical electronics, computers, and television sets. In terms of the chemical pollutants that accompany most alternative methods of providing light, heat (both for warmth and for cooking), locomotion, communication, or mechanical power, electricity is by far the cleanest. If anything, the major danger to this country comes from not relying still more on electrical power — through greater use of electric railroads instead of cars, buses, and trucks, for example, or even through using telecommunication to replace travel.

1

The Nature of Low-Frequency Electromagnetic Fields

A major triumph of nineteenth-century physics was the development by James Clerk Maxwell of a set of unified equations to describe the time and spatial dependence of electromagnetic fields (Maxwell 1873). These equations give precise quantitative agreement with observed classical phenomena over an enormous range in frequency — certainly from 0 (direct current, DC) to more than 10^{15} Hz (optical frequencies).[1] For atomic dimensions and frequencies comparable to atomic or molecular transitions, a satisfactory theory requires the combination of Maxwell's equations with quantum theory. But for the description of extremely low-frequency (ELF) effects at dimensions comparable to or larger than about 1 micron (μm, 10^{-6} m) and characteristic of the dimensions in cell biology, the classical form of Maxwell's equations should be reliable, although not all problems involving the interaction of ELF fields with the human body may be easily solved on an a priori basis.

Frequency Spectrum of Electromagnetic Sources

At present, coherent sources of electromagnetic radiation (sources with well-defined frequency and phase) have been developed over the range from DC to the vacuum ultraviolet without any appreciable gaps (fig. 1.1). Frequencies are specified in terms of Hz with Greek prefixes, according to the conventions shown in table 1.1.

The frequency regions emphasized in this book are indicated in figure 1.1. The first band (15–180 Hz) includes frequencies used in some electrified rail systems in Europe (16⅔ Hz) and the United States (25 Hz) (Shedd 1974), as well as the first three harmonics of the common power-line frequencies in both the United States (60 Hz) and most other countries (50 Hz). The second band (10–30 kHz) includes the fundamental horizontal

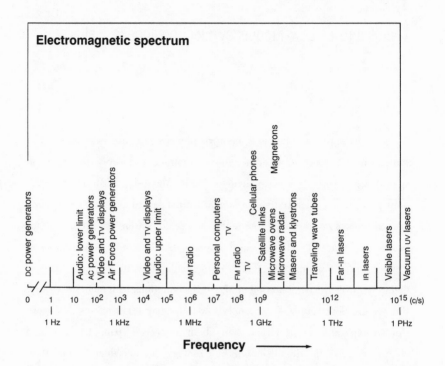

Figure 1.1. Schematic indication of the frequency range over which electromagnetic power generators have been produced. *Note:* The frequency bands with which this book is concerned are 15 to 180 Hz and 10 to 30 kHz.

Table 1.1. Nomenclature for Frequency and Time Interval Measurement

Frequency (f)	Written	Meaning	Period ($= 1/f$)	Written
0 hertz	DC	Direct current		
	AC	Alternating current		
1 hertz	1 Hz	1 cycle/second	1 second	1 s
1 kilohertz	1 kHz	10^3 cycles/sec	1 millisecond	1 ms
1 megahertz	1 MHz	10^6 cycles/sec	1 microsecond	1 μs
1 gigahertz	1 GHz	10^9 cycles/sec	1 nanosecond	1 ns
1 terahertz	1 THz	10^{12} cycles/sec	1 picosecond	1 ps
1 petahertz	1 PHz	10^{15} cycles/sec	1 femtosecond	1 fs
1 exahertz	1 EHz	10^{18} cycles/sec	1 attosecond	1 as

scanning frequency component (typically, about 15.75 kHz) of most television sets and many video display terminals used on personal computers. The power frequency adopted by the U. S. Air Force (400 Hz) falls in the gap between the two bands; although not specifically included here, problems at 400 Hz should not differ greatly from those in the lower band. By international convention, the frequency range of 30–300 Hz is described as extremely low frequency and the range of 3–30 kHz as very low frequency (VLF).

Spectra of Pulsed and Periodic Waveforms

Strictly speaking, even a source that varies sinusoidally has no precisely defined frequency, unless it is turned on for an infinite length of time. It follows generally from considerations based on Fourier analysis that the full frequency width Δf at half-maximum intensity of the power spectrum of a sinusoidal oscillation, which is turned on only for time T seconds, is given by

$$\Delta f \approx 1/T \text{ Hz.} \tag{1}$$

Figure 1.2 illustrates the power spectrum of a short sinusoidal pulse. As T becomes large compared with the period $1/f_0$ of the sinewave and many cycles of the sinewave are included, the spectral width (central peak in the bottom half of the figure) contracts to a very narrow distribution about f_0. If, however, $T \ll 1/f_0$, the energy is spread over a large spectral range. The width Δf of the central lobe increases with $1/T$, and a number of satellite peaks occur that spread outward from the central maximum at f_0 in a manner equivalent to the familiar single-slit diffraction pattern in optics. For extremely short pulses, the spectrum may actually spread over an amount larger than the frequency f_0 of the original sinewave.

Still more complicated results occur when the field is turned on and off periodically. Here, the waveform may be expressed as a discrete Fourier series in the fundamental frequency f_0 of the periodic wave by

$$F(t) = \sum_{n}^{\infty} C_n \sin (n2\pi f_0 t + \phi_n), \tag{2}$$

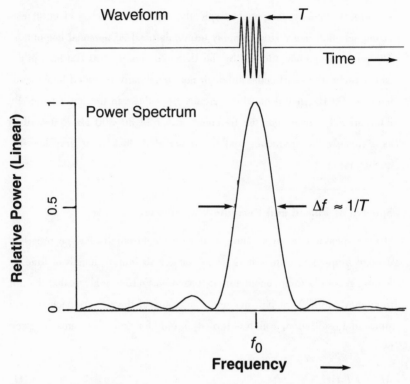

Figure 1.2. Short pulse of sinewaves (top) and its power spectrum (bottom)

where C_n is the amplitude of the nth harmonic and $n = 1, 2, 3, \ldots$ (For the purposes of the present study, we shall ignore the DC term, for which $n = 0$.) Here, ϕ_n is the phase angle for the nth harmonic, and the power spectrum is proportional to $|C_n|^2$ at frequency nf_0.

Because I deal in this book with both electric and magnetic fields and in particular with electric fields induced by the Faraday effect from time-varying magnetic fields, the power spectrum of such induced electric fields can be weighted substantially toward higher frequencies over those found in the original magnetic-field spectrum. The induced electric field in the Faraday effect is proportional to the time rate of change of the magnetic flux. Hence, if a periodic magnetic-field waveform is given by $F(t)$ in the above Fourier series, the electric field induced by it is proportional to

$$\frac{\mathrm{d}F}{\mathrm{d}t} = 2\pi f_0 \sum_{n}^{\infty} nC_n \cos\left(n2\pi f_0 t + \phi_n\right). \tag{3}$$

The relative power spectrum of the induced electric field is then proportional to $|n\, C_n|^2$ where C_n is the nth harmonic coefficient for the original periodic magnetic field. This effect can be of major importance in determining the spectral distribution of applied fields in both controlled studies and common environments.

An example taken from Bassett et al. (1981) provides useful insight. Application of periodic ELF magnetic fields is believed to have a beneficial effect on healing bone fractures, presumably through the action of induced electric fields arising from the Faraday effect inside the fractured bone. A periodic current waveform is applied through Helmholtz coils placed around the injury. According to Bassett and coworkers, a generally used configuration consists of a burst of about twenty constant-amplitude pulses of 200-μs duration and 50-μs spacing lasting almost 5 ms, applied periodically at a 15-Hz repetition frequency. Two cycles of this magnetic-induction field waveform are shown at the top of figure 1.3.

The relative power spectrum for this periodic waveform obtained by Fourier analysis is shown at the middle of the figure. (The spectra were computed by the methods discussed in Bennett [1976], chap. 7.) Although the peak alternating-current component is actually at 15 Hz, the spectrum exhibits a number of strong features in addition to the harmonics of the 15-Hz repetition frequency. The large peaks at 4, 8, 12, and 16 kHz are harmonics of the rectangular pulses of 250-μs period in the pulse burst. The widths of these harmonics are determined by the 5-ms duration of the pulse burst within each 1/15-s cycle in accordance with eq. (1). Finally, the many smaller peaks between the major peaks are analogous to diffraction fringes of the sort obtained optically from multiple-slit interference patterns.

As shown at the bottom of the figure, the spectral distribution is skewed to higher frequencies for components of the electric field induced through the time derivative of the magnetic flux (dB/dt). Here, the maximum spectral components occur in the range from about 8 to 12 kHz and are roughly 30 dB higher than the power at the 15-Hz repetition rate. Obviously, neither the spectrum for $B(t)$ nor for $dB(t)/dt$ is simply confined to the region of 15 Hz or even to a few harmonics of 15 Hz. The actual spectrum of power

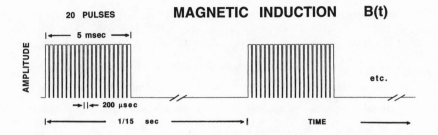

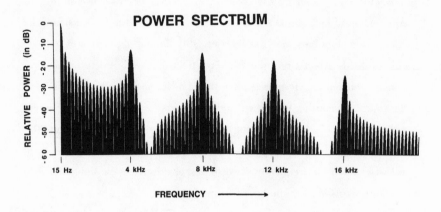

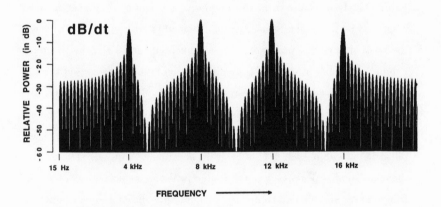

Figure 1.3. Periodic waveform (top) used for fractured-bone therapy. Relative power spectra of the magnetic field (middle) and induced electric field (bottom) from dB/dt are also shown. *Note:* $B_\omega(15 \text{ Hz}) = 0.1189 \, B_{max}(t)$.

dissipated in the bone would of course differ still further because of the frequency dependence of the bone conductivity. The basic point of this illustration is that the main frequency components of the currents induced in the body can differ enormously from the fundamental repetition frequency when the externally applied field is nonsinusoidal.[2] Effects of this type may well be responsible for inconsistencies in the literature regarding the existence of "frequency windows" in the biological effects of ELF fields. Of course, if the applied field is purely sinusoidal, this source of confusion does not arise.

Health threats from electromagnetic fields might conceivably be more related to field amplitude than to field power. For classical (that is, nonquantum) phenomena, power expended (or intensity) is generally proportional to the square of the field amplitude. Relative power measurements are often specified in decibels. Then, the power ratio is related to the corresponding amplitude ratio by the convention that the number of decibels is given by

$$10 \log_{10}(\text{power ratio}) = 20 \log_{10}(\text{amplitude ratio}) \quad [\text{dB}]. \tag{4}$$

AM and FM Spectra

Finally, it is important to comment on the spectral distribution obtained from low-frequency modulation of electromagnetic radiation at radio and microwave frequencies. Such modulation is sometimes erroneously described as producing ELF effects. The spectra obtained from amplitude and frequency modulations (AM and FM) differ in detail and are discussed separately.

The frequency spectrum occupied by an amplitude-modulated radio-frequency (rf) wave is determined by the bandwidth of the modulation signal. When the amplitude A of an rf sinewave at a carrier frequency ω_c is modulated at a low-frequency ω_m so that

$$A(t) = [A_0 + A_m \cos (\omega_m t)] \sin (\omega_c t), \tag{5}$$

the resultant time variation of the field can be rewritten as

$$A_0 \sin (\omega_c t) + 0.5 A_m \sin [(\omega_c + \omega_m)t] + 0.5 A_m \sin [(\omega_c - \omega_m)t]. \tag{6}$$

Hence, the spectrum consists of the original carrier frequency at ω_c, plus upper and lower sidebands at $(\omega_c + \omega_m)$ and $(\omega_c - \omega_m)$. If the low-frequency wave has components extending over a bandwidth $\Delta\omega_m$, the rf spectrum occupied by the wave extends from $(\omega_c - \Delta\omega_m)$ to $(\omega_c + \Delta\omega_m)$. If $(\omega_c - \Delta\omega_m) \gg \omega_m$, there is no spectral component in the vicinity of ω_m. To take an example, in spite of occasional comments to the contrary found in the literature, if a 450-MHz radio wave is squarewave-modulated at 15 Hz, there is no significant spectral energy at 15 Hz. Even if the squarewave had a rise and fall time as short as 1 nanosecond (ns), the power at 15 Hz would be down by at least 150 dB from the power at the carrier frequency. Of course, there may be important biological effects from such electromagnetic fields at the radio frequency, but they cannot be regarded as direct effects of ELF fields. There is some evidence, however, that nonlinearities in biological material may result in the detection of ELF modulation on radio frequency signals. (See chapter 6.)

Frequency modulation differs inherently from amplitude modulation in that the spectral bandwidth occupied at radiofrequencies increases with modulation amplitude. When the frequency of an rf carrier wave is modulated sinusoidally, the radio wave takes on the form

$$E(t) = A_0 \sin [\omega_c t + A_m \sin (\omega_m t)], \tag{7}$$

where the various terms are defined as they were in the AM case. This waveform may be rewritten as a series of sidebands at frequencies $(\omega_c + n\omega_m)$ and $(\omega_c - n\omega_m)$ with amplitudes J_n, where the J_n are Bessel functions of real argument of order n, and $n = 0, 1, 2, 3, \ldots$. [3]

The result is a series of sidebands about the carrier frequency from sinewave modulation similar to those found in the AM case, except that the amplitudes of the sidebands vary in a nonlinear way with the modulation amplitude A_m. For $A_m = 0$, the only nonzero amplitude is the one for the carrier frequency itself. As A_m increases, the peak amplitudes shift toward the outer sidebands. Although the frequency extent of the sidebands is inherently greater for the same ratio of A_m/A_0 than in the amplitude-modulation case, the values of J_n fall off exponentially as n moves out from the peak sideband amplitude. Hence, if ω_m corresponds to an ELF signal and ω_c is in the rf range, there will be a negligible ELF component in the spectrum once again. The ELF contribution is not quite as small as in the AM case, however.

Field Equations and Boundary Conditions

Although the solution of Maxwell's equations (table 1.2)[4] can be quite formidable when the electromagnetic wavelengths are comparable to or smaller than the physical dimensions of the material objects involved, there is an enormous simplification at the opposite extreme. Even for the highest frequencies of concern here (30 kHz), the wavelengths λ ($= c/f$, where c is the velocity of light and f is the frequency) of the radiation are infinite, for all practical purposes, compared with the dimensions of the objects involved. For example, the free-space wavelengths range from about 10 km at 30 kHz to 30,000 km at 10 Hz. Under these conditions, we can get extremely good solutions to Maxwell's equations merely by solving the equivalent static problem and by then multiplying those solutions for the electric or magnetic field by sinusoidal time-dependent factors at the frequency of concern. Thus, for the purpose of determining the spatial variation of the fields, the time derivatives of the fields in table 1.2 can be set equal to zero. In this limit, the electric field $\mathbf{E}$ and the magnetic field $\mathbf{H}$ decouple, and we need only the much simpler solutions to the separate, static equations for $\mathbf{E}$ and $\mathbf{H}$. One small exception to this situation occurs in the calculation of electric fields through the Faraday effect from slowly varying magnetic fields.

Not only are Maxwell's equations well tested over a wide range of frequencies and dimensions, but solutions to them are much easier to obtain

Table 1.2. Classical Electromagnetic Field Equations

Maxwell's Equations	Most Materials	Continuity of Change
$\nabla \times \mathbf{E} = -\dfrac{\partial \mathbf{B}}{\partial t}$		
	$\mathbf{D} = \epsilon \mathbf{E}$	
$\nabla \cdot \mathbf{D} = \rho$		
	$\mathbf{J} = \sigma \mathbf{E}$	$\nabla \cdot \mathbf{J} + \dfrac{\partial \rho}{\partial t} = 0$
$\nabla \times \mathbf{H} = \mathbf{J} + \dfrac{\partial \mathbf{D}}{\partial t}$		
	$\mathbf{B} = \mu \mathbf{H}$	
$\nabla \cdot \mathbf{B} = 0$		

Note: The first column contains Maxwell's equations in their complete form. The second column contains relationships that hold for many isotropic "linear" materials, but not for all materials. The third column represents the continuity relation between charge and current. **E** is the electric field, **D** is the displacement, **H** is the magnetic field, and **B** is the magnetic induction. For a given material, ϵ is the permittivity, σ is the conductivity, and μ is the magnetic permeability. In the equation of continuity, **J** is the current flux and ρ is the free-charge density.

for the frequencies considered in this book than they would be at frequencies in excess of 100 MHz, for which the wavelengths are comparable to the dimensions of the human body.

Force on a Charged Particle in Electromagnetic Fields

The force **F** on a charged particle q in combined electric and magnetic fields is given by

$$\mathbf{F} = q(\mathbf{E} + \mathbf{v} \times \mathbf{B}) \quad [\text{newtons}], \tag{8}$$

where the charge q is in coulombs, the electric field **E** is in volts per meter, the velocity of the charged particle **v** is in meters per second, and the magnetic induction field **B** is in teslas (1 milligauss [mG] = 0.1 microtesla [μT]).[5] The first term is the force from the electric field, and the second is the force from the magnetic field. The full expression in eq. (8) is known as the "Lorentz force," and the equation is correct even when the charge is moving near the velocity of light. The force from the cross product (second term) is perpendicular to the plane of $\mathbf{v}$ and $\mathbf{B}$ and in the direction that a right-hand screw would advance when $\mathbf{v}$ is rotated into $\mathbf{B}$. The second term results in an effective, additional electric field (force per unit charge) of amount $\mathbf{v} \times \mathbf{B}$ when the charge moves through a magnetic field, and it is the basic principle behind electric power generators. In that case, the charges are constrained to electric wires moving through magnetic fields. As an example, the effective DC electric field acting in a person traveling across the United States in an airplane at 500 miles per hour (223.5 m/s) is about 0.010 V/m.

The work done on a charge moving through the force is entirely produced by the first term in eq. (8) (the "classical" electric field), because the force from the magnetic field is perpendicular to the velocity of the particle. Hence, if a particle starting from rest accelerates through a potential V (the integral of $\mathbf{E} \cdot d\mathbf{r}$ over the path), the velocity acquired by the charge is given by

$$\tfrac{1}{2}m_0 v^2 = qV \quad \text{[joules]}, \tag{9}$$

where m_0 is the rest mass of the particle; and the equation holds in the limit that $v^2 \ll c^2$. For electrons and other elementary particles, the energy acquired by the charge in moving through 1 volt is known as 1 electron volt: $1 \text{ eV} = 1.6 \times 10^{-19}$ joules (J).

Because some large modern television sets use mildly relativistic electron beams, I shall note the modifications required in that limit. There, it is a useful calculational aid to introduce the notion of a "relativistic mass" m, defined in respect to the rest mass m_0 by $m/m_0 = [1 - (v/c)^2]^{-1/2}$. The total energy acquired by a charge in falling through a potential V is then given by $(m - m_0)c^2 = qV$; and the acceleration produced by the magnetic field is $q\mathbf{v} \times \mathbf{B}/m$, but with the velocity-dependent "relativistic mass" in the denominator.

Boundary Conditions between Media

Maxwell's equations result in well-defined boundary conditions at the surface between different media, as follows.[6]

The tangential components of the electric field are continuous:

$$E_{1\mathrm{tan}} = E_{2\mathrm{tan}}. \tag{10}$$

The discontinuity in the normal component of the displacement vector equals the surface charge density at the boundary:

$$D_{2\mathrm{normal}} - D_{1\mathrm{normal}} = \rho_s. \tag{11}$$

Provided that neither medium has infinite conductivity and there is no surface current at the boundary, the tangential components of the magnetic field are continuous:

$$H_{1\mathrm{tan}} = H_{2\mathrm{tan}}. \tag{12}$$

The normal components of the magnetic induction field are continuous:

$$B_{1\mathrm{normal}} = B_{2\mathrm{normal}}. \tag{13}$$

Finally, if the two materials have nonzero conductivity, and the fields vary slowly, the equation of continuity implies that surface-charge density ρ_s can build up at the interface such that

$$J_{2\mathrm{normal}} - J_{1\mathrm{normal}} = -\mathrm{d}\rho_s/\mathrm{d}t. \tag{14}$$

These boundary conditions determine the fields that exist inside biological tissue in the presence of externally applied fields.

Near Field versus Far Field

Another important consequence of the enormously long wavelengths involved in the low-frequency region is that far-field (radiation) effects are of no practical importance. By far field is meant $r \gg \lambda$ or that the observer is at a distance from the source that is very much greater than the wavelength.

(At 60 Hz, one would have to be much more than 5,000 km from the source.) If biological damage arises from exposure to devices generating such low-frequency fields, it is due entirely to the near-field ($r \ll \lambda$) characteristics of the problem. Indeed, the primary fields of importance produced by long, straight parallel 60-Hz power lines simply do not result in any significant far fields at all. Because the electric- and magnetic-field lines fall in a plane perpendicular to the wire, the Poynting vector $\mathbf{E} \times \mathbf{H}$, which determines the power flow, is in a direction parallel to the wire. Thus, the flow of power is along the transmission lines and is not radiated away perpendicularly. One could of course build a special antenna to radiate power at 60 Hz. To do so efficiently, however, would require antenna dimensions of the order of $\lambda/2 \approx 2,500$ km.

Nonionizing Nature of the Fields

Table 1.3. Physical Constants

Constant	Value
Boltzmann's constant	$k = 1.3807 \times 10^{-23}$ joule/K
kT at body temperature (310 K)	4.28×10^{-21} joule $= 0.0268$ eV
Planck's constant	$h = 6.6262 \times 10^{-34}$ joule-s
Permeability of free space	$\mu_0 = 4\pi \times 10^{-7}$ henry/meter
Permittivity of free space	$\epsilon_0 = 8.854 \times 10^{-12}$ farad/meter
Electrical energy	1 eV $= 1.6022 \times 10^{-19}$ joule
Charge of the electron	$e = 1.6022 \times 10^{-19}$ coulomb
Mass of the electron	$m_e = 0.91095 \times 10^{-30}$ kilogram
Atomic mass unit	1 amu $= 1.6606 \times 10^{-27}$ kilogram
Speed of light in vacuum	$c = 2.9979 \times 10^8$ m/s

The energy $h\nu$ per photon at the frequencies examined here is negligible compared with both molecular binding energies and thermal energies. At the upper end of the frequency range in this study (30 kHz),

$$h\nu \approx 2 \times 10^{-29} \text{ joule} \approx 1.24 \times 10^{-10} \text{eV} \approx 4.8 \times 10^{-9} kT, \tag{15}$$

Table 1.4. Conversion of SI to Gaussian Units

Symbol	Entity	SI (rationalized, mks)	Gaussian (electrostatic, cgs)
l	Length	1 meter	100 cm
m	Mass	1 kg	1,000 gm
t	Time	1 s	1 s
F	Force	1 newton	10^5 dynes
P	Power	1 watt	10^7 ergs/s
C	Capacity	1 farad	$1/(4\pi\epsilon_0) \approx 9 \times 10^{11}$ cm
q	Charge	1 coulomb	3×10^9 statcoulomb (esu)
ρ	Charge density	1 coulomb/m^3	3×10^3 statcoulomb/cm^3
σ	Conductivity	1 siemen/m	9×10^9 s^{-1}
I	Current	1 ampere	3×10^9 statampere
J	Current density	1 ampere/m^2	3×10^5 statampere/cm^2
D	Displacement	1 coulomb/m^2	$12\pi \times 10^5$ statcoulomb/cm^2
E	Electric field	1 volt/meter	$1/3 \times 10^{-4}$ statvolt/cm
$V, \mathscr{E}$	Electromotive force	1 volt	1/300 statvolt
W	Energy	1 joule	10^7 ergs
f, v	Frequency	1 hertz	1 cycle/sec
Z	Impedance	1 ohm	$4\pi\epsilon_0 \approx 1/9 \times 10^{-11}$ s/cm
L	Inductance	1 henry	$4\pi\epsilon_0 \approx 1/9 \times 10^{-11}$ s^2/cm
Φ	Magnetic flux	1 weber	1/300 esu($= 10^8$ maxwells)
B	Magnetic induction	1 tesla	10^4 gauss (1 mG = 0.1 μT)
H	Magnetic field	1 amp-turn/m	$4\pi \times 10^{-3}$ oersted
μ	Magnetic moment	1 amp-m^2	10^3 oersted-cm^3
M	Magnetization	1 amp-turn/m	10^{-3} oersted
R	Resistance	1 ohm	$4\pi\epsilon_0 \approx 1/9 \times 10^{-11}$ s
ρ	Resistivity	1 ohm-m	$1/9 \times 10^{-9}$ s
μ_0	Permeability	1 henry/m	$1/4\pi \times 10^7$
ϵ_0	Permittivity	1 farad/m	$36\pi \times 10^9$
P	Polarization	1 coulomb/m^2	3×10^5 statcoulomb/cm^2
p	Pole strength	1 weber	$1/1200\pi$ esu($= 10^8/4\pi$ emu)
V, ϕ	Potential	1 volt	1/300 statvolt

Source: Data from Book (1987), Smyth and Ufford (1939), and Jackson (1975)
Note: Equality is assumed across the two columns.

whereas at 60-Hz power-line frequencies and a body temperature $T \approx$ 310 K

$$h\nu \approx 4 \times 10^{-32} \text{ joule} \approx 2.5 \times 10^{-13} \text{eV} \approx 9.3 \times 10^{-12} kT, \qquad (16)$$

so that $h\nu \ll kT$. Hence, any experiment designed to detect single-photon events at these frequencies would be overwhelmed by thermal noise.

Because biological molecules are fairly stable at such temperatures, their binding energies are much greater than the energy of a single photon. Consequently, single-photon ionization or excitation of such molecules (which is important for $h\nu \gg kT$) from these low-frequency sources must be negligible. Although ionization can occur in high enough electric fields through classical (multiphoton) processes, "radiation" (even if it existed) at the frequencies discussed in this book is definitely "nonionizing" in the usual medical radiation-physics sense of the term. Some useful constants are summarized in table 1.3 in SI (that is, meter-kilogram-second, or mks) units; the conversion to the older cgs units is given in table 1.4. With a few exceptions (such as the use of gauss for magnetic induction) or as otherwise noted, SI units are used throughout.

2

Sources of Low-Frequency Fields

Power utilities and the Environmental Protection Agency have published some measurements of electric and magnetic field strengths found near high-voltage power lines. The Bonneville Power Administration provided the measurements shown in table 2.1. As the administration explained, long-distance transmission from hydroelectric-power sources takes place over 115- to 500-kV lines along tracts where the utility has a restricted right-of-way extending about 50 to 65 ft. (15 to 20 m) laterally from the line. Near the user location, substation transformers reduce the voltage to feed 12-kV distribution lines,[1] which are carried over street poles (or cables buried in underground conduits) throughout urban areas and from which pole-mounted transformers (or corresponding underground units) provide shorter 240/120-V lines to residential and other users.

Another widely quoted study deals with electric fields measured under the Marysville 765-kV line, where the fields vary from about 3 to 10 kV/m near ground level (the line sags).[2]

Reports seem to concentrate either on high-tension power lines or on the remarkably small magnetic fields, on the order of a few mG to less than 0.1 mG, measured in or near home environments.[3] Apart from two semi-quantitative illustrations attributed to Florig et al. (1987) that have been reproduced many times, relatively little data seem to be available on the fields associated with the common 12-kV distribution lines in the urban environment.

For these reasons, and in order to gain more insight regarding the magnitudes and underlying sources of ELF fields in the common human environment, this chapter contains a review of simple methods of field computation. Even at very low frequencies, closed-form solutions to Maxwell's equations are difficult to obtain unless an unusual degree of symmetry

Table 2.1. Representative Electric- and Magnetic-Field Strengths at Ground Level near High-Tension Lines

	Distance from Line in Feet (m)				
	0 (0)	50 (15)	65 (20)	100 (30)	200 (61)
500-kV line					
Electric field (kV/m)	7.0		3	1	0.3
Magnetic field (mG)	70		25	12	3
230-kV line					
Electric field (kV/m)	2.0	1.5		0.3	0.05
Magnetic field (mG)	35	15		5	1
115-kV line					
Electric field (kV/m)	1.0	0.5		0.07	0.01
Magnetic field (mG)	20	5		1	0.3

Source: Data from Bonneville Power Administration (June 1989)

Note: Although the electric-field values are fairly independent of the load, the magnetic fields increase with the current through the wires

is involved. Yet some limiting cases can be solved, providing insight regarding the nature of the general problem. In many instances these special cases yield a good estimate for the fields with which this book is concerned.

Magnetic Fields

Because the initial report by Wertheimer and Leeper (1979) of a possible link between childhood leukemia and magnetic fields from power lines stimulated the controversy over possible deleterious health effects from ELF electromagnetic fields, it is especially important to consider methods for accurately determining such fields.

Magnetic Fields from Straight Wires

First, consider the magnetic field from an infinitely long wire carrying a total current I. It can be shown by the application of Stokes's theorem to the third of Maxwell's equations in table 1.2 that the magnetic-field lines

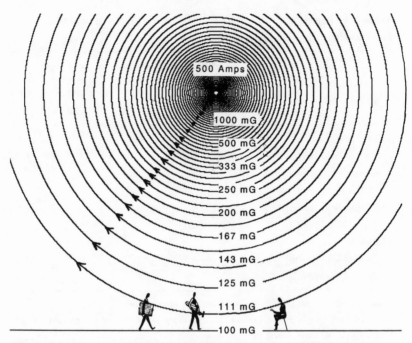

Figure 2.1. Magnetic induction field produced by a long, straight wire. *Note:* The field lines and strengths assume that a 500-A current flows into the diagram 10 m above the ground. The total field magnitudes are tabulated (in mG) at 1-m intervals above the ground, directly under the wire. (For comparison, subsequent figures have been computed for the same current per wire and height above ground.)

at a distance *r* from the wire are circular and the field is of a constant magnitude given by

$$H = I/2\pi r \quad \text{or} \quad B = \mu_0 I/2\pi r \quad \text{[SI units]} \tag{17}$$

or, more simply, with *B* in mG (but *I* in amperes and *r* in m),

$$B = 2I/r \quad \text{[mG]} \tag{18}$$

in free space (or in nonmagnetic materials). The field is in the direction that the fingers would curl around the wire if it were held in the right hand with the thumb aimed toward the current flow.

An example of the magnitudes of the magnetic induction field produced by a long, straight wire is shown in figure 2.1, assuming that the wire carries 500 A (going into the diagram) and is located 10 m above the ground. The fields directly under the wire are shown in mG in 1-m steps, starting at ground level. The human figures in this and in subsequent illustrations are intended to give scale. Although the height of power lines and the current flowing through them may vary, the numbers chosen in these illustrations are representative of the upper limit one might find in common situations. The magnetic fields passing through people directly under the power line in this instance are on the order of 60 mG, or about one tenth the magnitude of the earth's static magnetic field. In this type of plot, the direction of the field is given by lines and arrows. It is a useful convention to vary the spacing between the lines inversely with the magnitude of the magnetic flux. Because B varies as $1/r$ (eq. [18]), the lowest fields occur farthest from the wire where the spacing between the field lines is largest, and vice versa. The figure corresponds either to the static field from a constant current flowing into the diagram or to the magnetic field from an alternating current at one instant in time. If the current is slowly varying with time sinusoidally, the field lines alternately contract and expand in phase with it and reverse their direction at the power-line frequency. The magnitude of the field at any instant is proportional to the current. Alternating currents are usually specified in terms of root-mean-square (rms) values, rather than peak amplitudes, through the relation

$$I_{rms} = I_{peak}/\sqrt{2} \approx 0.707\, I_{peak}. \tag{19}$$

Hence, for single-phase lines, if rms values of the current are specified, the application of eq. (18) results in rms values of the magnetic flux.

Parallel Single-Phase Lines

For the single-wire example shown in figure 2.1, it is tacitly assumed that the wire carrying the current returning to the power source is infinitely far away. In reality, there must always be a return wire, and in practice it is usually close to the other wire. The result obtained above for a single wire must therefore be modified for parallel transmission wires. As may generally be seen by the application of Stokes's theorem (or from Ampere's law), the near presence of the return wire automatically reduces the magnetic field at distances from the wires that are large compared with their spacing. If the currents flowing in opposite directions in two parallel wires are equal, the line integral of the field taken around the pair is zero, because no net current is enclosed within the loop. Closer to the pair of wires, there is only partial cancellation of the field. It is of interest to examine what happens to the field near the wires in a few representative cases.

We may easily extend the single-wire solution to determine the magnetic field from two parallel wires, or any number of parallel wires, by using the principle of superposition. Because Maxwell's equations are linear, any linear combination of separate solutions to them must also be a solution. But it is the *vector* sum of the separate fields from the separate wires that concerns us here. Let us rewrite eq. (18) treating the current $\mathbf{I}$ as a vector. The magnetic flux at a point $\mathbf{P}$ from a group of n parallel wires is then given by

$$\mathbf{B} = C \sum_{i=1}^{n} (\mathbf{I}_i \times \mathbf{r}_i)/r_i^2, \tag{20}$$

where $\mathbf{r}_i = \mathbf{P} - \mathbf{W}_i$. The constant $C = 2$ mG-m/A; the vector $\mathbf{W}_i$ points to the location of the ith wire; and $\mathbf{P}$ is a vector defining a particular point in space from a common origin in a plane perpendicular to the wires.

Figures 2.2 and 2.3 were computed from equation (20) for wires 2 m apart with the midpoint 10 m above the ground, carrying 500 A in opposite directions. The current and height are the same as in figure 2.1 for the single-wire example, but there is substantial cancellation of the magnetic field near the ground level in comparison with the single-wire line. The cancellation at ground level does not differ significantly between the two double-wire configurations. Power companies often use the vertical geometry with much closer spacing on the low-voltage (240/120 V), high-current distribution lines from power transformers to residential houses; the horizontal configuration is more frequently used in high-voltage, low-current lines.

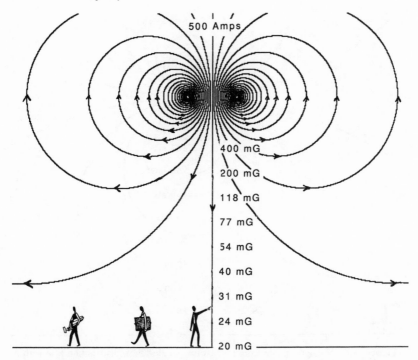

Figure 2.2. Magnetic fields from a two-wire, single-phase distribution line 10 m above ground with a horizontal spacing of 2 m. *Note:* 500 A flows into the diagram at the left and out at the right. (Two-wire single-phase lines are mainly used in rural areas or on short runs, as opposed to the more common three-phase distribution lines used in urban areas.)

From the density of the flux lines in figures 2.2 and 2.3, it is evident that the near cancellation of the magnetic field at large distances is compensated for by a large increase in the field between the two wires. In general, the smaller the separation between the two wires, the more exaggerated the difference. Although the field from a single long, straight wire falls off as $1/r$ (where r is the distance from the wire), the field from two (or more) wires carrying equal and opposite currents in both directions falls off as $1/r^2$ when the distances are large compared with the wire separation.

Magnetic Fields from Electrified Railroads

Three-wire single-phase power lines are used in electrified trains, trolley cars, and subways. Each of two parallel-current return lines (the rails,

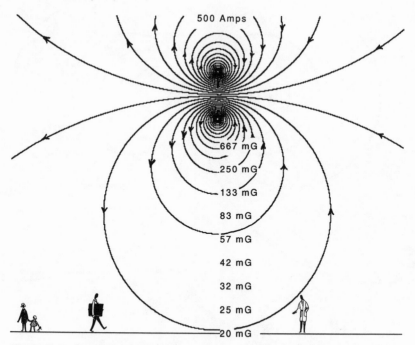

Figure 2.3. Magnetic fields from a two-wire, single-phase distribution line with a midpoint at 10 m above ground and a vertical spacing of 2 m. *Note:* 500 A flows into the diagram at the top and out at the bottom.

which are about 5 ft. [1.52 m] apart) laid at ground level carries half the current. The high-voltage feed comes either from a single overhead line or from an insulated third rail. (New York City subways and the San Francisco BART lines use DC transmission on the third rail and are excluded from this study.) At least eight railroads in the United States use 11,000-V AC power lines to supply engines with continuous ratings of 4,000–6,800 horsepower. In many instances, step-down transformers and DC rectifiers are used within the locomotive to drive DC motors. The overhead power line must supply about 300–500-A AC to run just one engine. Although 60-Hz power lines with rectification inside the engine and DC motors are commonly used today, some of the earlier lines (such as the New York, New Haven, and Hartford Railroad) formerly used a line frequency of 25 Hz to drive AC motors directly (Partridge 1967). Twenty-five Hz is still used on the Pennsylvania branch of Amtrak.

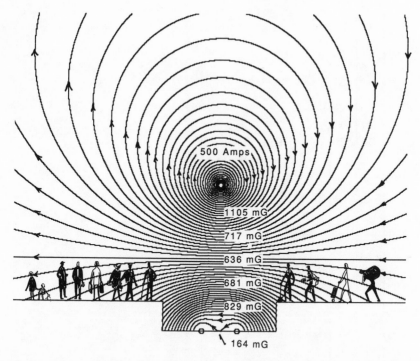

500 Amps

1105 mG

717 mG

636 mG

681 mG

829 mG

164 mG

Figure 2.4. Magnetic fields from electrified trains. *Note:* 500 A flows into the diagram in the overhead wire and is returned in equal amounts by the two rails.

Figure 2.4 illustrates the magnetic-field lines produced by a 500-A current flowing in the overhead wire and being returned in equal amounts by the two rails. The rms field magnitudes are shown in table 2.2. These data correspond approximately to field measurements during maximum bursts of acceleration on the 25-Hz branch of Amtrak between Washington, D.C., and New York City. People need not be on the train to experience the field as long as they are between the power source and the engine. In this geometry, the gap between the opposing currents includes the level where people would be located, both on the station platform and inside the train. There they would be exposed to fields substantially stronger than the ones illustrated earlier. Here, the magnetic fields are of about the same magnitude as the earth's static magnetic field. Although considerable shielding from the magnetic field would result inside an iron or steel train, relatively little distortion of these fields would be expected in a train made

Table 2.2. Worst-Case RMS Magnetic Fields near an Electrified Railroad (mG)

y (m)	0	2	4	6	8	10	20	40	60	80
							x (m)			
6	10,164	483	213	121	77	53	15	4	2	1
5	1,105	497	232	129	81	55	15	4	2	1
4	717	471	242	135	83	56	15	4	2	1
3	636	460	245	137	84	56	15	4	2	1
2	681	483	245	135	83	56	15	4	2	1
1	829	543	237	130	81	54	15	4	2	1
0	164	556	217	121	76	52	15	4	2	1

Note: The overhead wire is $\approx$ 6.1 m above the track; y is the height above the track, and x is the horizontal distance from the wire. It is assumed that a current of 500 A rms flows through the overhead wire and is returned in equal amounts by the rails.

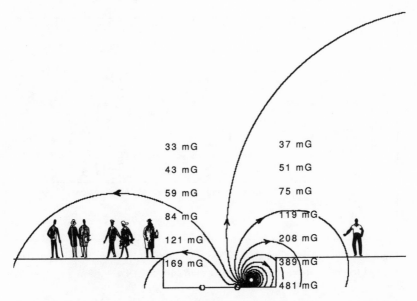

33 mG

37 mG

43 mG

51 mG

59 mG

75 mG

84 mG

119 mG

121 mG

208 mG

169 mG

389 mG

481 mG

Figure 2.5. Magnetic fields from subway trains. *Note:* 500 A flows into the diagram at the right through the "third rail" and is returned in equal amounts by the two other tracks.

of nonmagnetic stainless steel or magnesium-aluminum alloys. (Flat structural sections made of regular steel at the bottom of the cars would have relatively little effect on the horizontal field lines at seat level.) In the case of metal outer walls, electrical conductivity of the wall material would result in some minor reduction of AC magnetic fields because of eddy currents induced by the Faraday effect. (These induced, opposing fields increase in proportion to the line frequency.) The metal conducting body of the train would provide quite substantial shielding from external electric fields, however.

In figure 2.5 we see a similar set of magnetic-field lines, calculated for a subway environment incorporating a "third rail" near ground level. For purposes of comparison, the fields were again computed for a 500-A current. That value of the current is unrealistic for subways in the United States, however, for they all operate on relatively low-voltage, high-current DC.

In table 2.2 we saw the rms magnetic fields that would be produced from the overhead-wire electrified train, assuming a current of 500-A rms. In

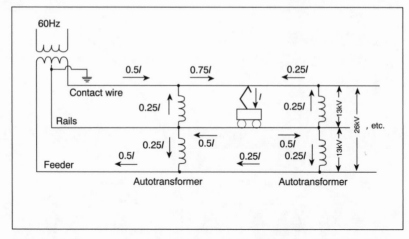

Figure 2.6 Wiring configuration for the post–1986 60-Hz New York–New Haven railroad (courtesy Metro-North commuter railroad)

this "worst-case" wiring geometry, the field does not drop off to 1 mG until about 80 m from the track. The calculation assumes that there is no significant source of magnetic shielding.

Fields from Modified Wiring on Electrified Railroads

The magnetic fields in figure 2.4 and table 2.2 are proportional to the total current in the overhead wire and represent worst-case limits for wiring configurations used earlier in the twentieth century. Some rail lines use more elaborate wiring configurations of the type shown in figure 2.6, which reduce the magnetic fields significantly. The specific system shown in the figure was adopted by the Metro-North commuter railroad in September 1986 when it converted from 25 to 60 Hz on the New York–New Haven route.[4] This lower-field configuration was partly adopted to minimize 60-Hz interference with telephone communication lines.

Single-phase 60-Hz power is supplied from a 26-kV secondary winding of a transformer that has a grounded center tap connected to the rails. One outer terminal of the secondary winding is connected to the overhead contact wire (trolley line), which is 13 kV with respect to the rails. The other outer terminal provides an out-of-phase 13-kV supply to a feeder line mounted at the edge of the track above the trolley wire. At intervals of a few miles along the route, the feeder line provides power to the trolley wire and the rails through a series of center-tapped autotransformers. The current going to the engine connected between the trolley wire and the rails is split up into several branches.

The configuration reduces the current flowing through the rails and the contact wire by a factor of about two over the earlier wiring system assumed in figure 2.4. There is additional magnetic-field cancellation because of the current flow parallel to the track in the out-of-phase feeder line. The higher voltage supplied also permits drawing lower current for the same engine power. Table 2.3 represents the calculated field magnitudes for the wiring configuration in figure 2.6, assuming the engine power to be the same as in table 2.2.

The situation is more complicated when multiple-unit commuter trains are used. Again using the Metro-North commuter railroad as an example, one of these self-propelled nonmagnetic stainless-steel cars draws about 80 A for maximum acceleration and about 40 A for steady, rapid travel. But the current is conducted from the pantograph in opposite directions along the car and down to the rails at each end, providing further reduction of the field inside the car. Commuter trains consist of three to ten of these cars, several of which may have their pantographs connected to the trolley wire. Thus, maximum currents could reach 800 A, although typical values for travel on a commuter express are mostly below 300 A.

Spot measurements of the magnetic fields from both Metro-North commuter trains and Amtrak express trains were made on this line using a three-coil magnetometer (a Field Star 1000, made by the Dexsil Corporation). This microprocessor-controlled unit has a range from about 0.04 mG to 1,000 mG and can display the resultant magnitude from simultaneous measurements of all three field components at 1-s intervals.

In the first car of a four-car express Metro-North Metroliner from New Haven to Bridgeport, Connecticut, the fields at chest level varied during acceleration from about 90 to 145 mG, were typically in the range from

Table 2.3. RMS Magnetic Fields for the Wiring Configuration Shown in Figure 2.6 (mG)

y (m)	x (m)									
	0	2	4	6	8	10	20	40	60	80
Location a										
6	6,602	278	102	51	30	19	5	1	1	0
5	633	266	109	55	32	21	5	1	1	0
4	378	238	112	57	33	21	5	1	1	0
3	315	222	111	58	34	21	5	1	1	0
2	322	224	108	56	33	21	5	1	1	0
1	381	243	102	54	32	21	5	1	1	0
0	84	242	92	49	30	20	5	1	1	0
Location b										
6	2,342	170	97	60	40	28	8	2	1	1
5	339	181	102	63	41	28	8	2	1	1
4	254	181	106	64	42	29	8	2	1	1
3	245	186	108	65	42	29	8	2	1	1
2	277	203	109	64	41	28	8	2	1	1
1	348	235	107	61	40	27	8	2	1	1
0	60	248	100	57	38	26	8	2	1	1

Note: The overhead trolley wire was 6.1 m above the track; the out-of-phase feeder line was assumed to be directly above and 50 percent higher than the trolley wire; y is the height above the track and x is the horizontal distance away from the center of the track. The engine current was assumed to be 440 A. Location a is to the left of the engine, and location b to the right in figure 2.6. Including the two rails, currents flow in four parallel conductors.

about 40 to 60 mG throughout the trip, and went down to roughly 20 mG when the train was coasting or braking. The minimum field encountered was 1.6 mG when the train went through a steel trestle bridge. The fields on the station platform at Bridgeport during commuting hours were typically 20 to 30 mG, and from the fluxmeter readings I could easily detect an arriving train long before it was visible. (With such a meter, you don't need to put your ear on the rail!) The return trip on the third car of a six-car, very slow Metroliner local yielded much lower fields: about 30 to 90 mG during acceleration, with typical values of about 20 mG. Indeed, larger fields on the order of 50 mG were encountered within the local when express trains passed from the opposite direction. I obtained similar results on electric commuter trains between Philadelphia and Bryn Mawr, Pennsylvania, where I used a correction for the meter response at 25-Hz.

I recorded the time variation of the magnetic field at chest level at 2-s intervals while seated in the rear car of a nine-car Amtrak Metroliner pulled by an electric locomotive during an entire six-and-a-half-hour trip from Washington to New Haven. Data for the 25-Hz fields from Washington to New York are shown in figure 2.7a and those for the 60-Hz fields on the New York–New Haven leg of the trip are shown in figure 2.7b. As is evident from both figures, these trains are driven by short bursts of power during acceleration — where the peak fields are encountered — but they spend most of their time either coasting or receiving just enough power to overcome friction. The 25-Hz fields during the Washington to New York branch of the trip were the largest. There, peak fields of up to 646 mG were encountered inside the train at sporadic intervals of one or two minutes' duration, with an average value over the four-and-a-half-hour trip of 126 mG. The peak observed field agrees with the maximum calculated values shown in figure 2.4 and table 2.2. Because of the different wiring arrangement that we saw in figure 2.6, peak 60-Hz fields on the New York–New Haven run amounted to only about 300 mG, with an average value over the whole two-hour trip of about 35 mG. The peak 60-Hz field shown in figure 2.7b agrees well with the values calculated in table 2.3.

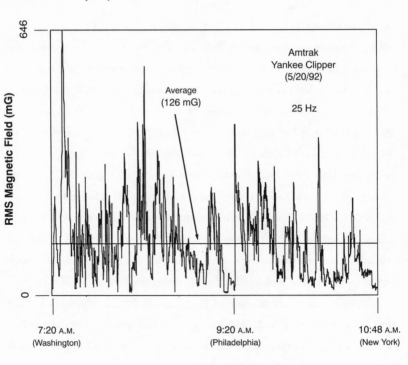

Figure 2.7a Time variation and average rms magnetic field at chest level for a person seated in the last car of a nine-car Amtrak Metroliner pulled by an electric locomotive, Washington to New York. *Note:* Measurements were made at 2-s intervals using a Dexsil Field Star 1000 three-coil magnetometer throughout the trip. (The locomotive and magnetometer batteries were changed in the New York station.)

As I shall discuss in detail later, any potential danger to health from such magnetic fields would arise primarily from electric fields induced in the body because of the Faraday effect. Such electric-field amplitudes occur in direct proportion to the frequency. Because the frequency ratio (60/25) roughly equals the field ratio between the data examined in figures 2.7a and 2.7b, there would be no significant difference in electric fields generated by the Faraday effect within the passengers on the two legs of the trip. The maximum fields in each case turn out to be smaller than thermal fields at the cell level by a factor of more than ten. These fields, however, do represent the maximum widely distributed magnetic fields encountered in the common environment studied here.

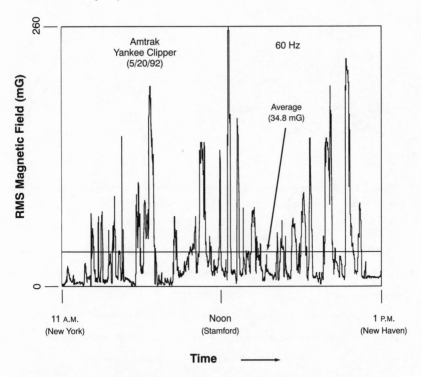

Figure 2.7b Time variation and average rms magnetic field at chest level for a person seated in the last car of a nine-car Amtrak Metroliner pulled by an electric locomotive, New York to New Haven. *Note:* Measurements were made at 2-s intervals using a Dexsil Field Star 1000 three-coil magnetometer throughout the trip. (The locomotive and magnetometer batteries were changed in the New York station.)

Magnetic Fields from Three-Phase Transmission Lines

Although a pair of wires carrying equal currents in opposite directions is commonly called a *single-phase* line, it could also be regarded as a two-phase system in which the phases differ by 180°. That distinction is useful for extending this discussion to three-phase lines.

For practical reasons — which involve the efficiency of AC generators and minimizing the loss from ohmic heating in the line — three-phase systems are generally used for long-distance power transmission (Robinson 1974, 595–597; Gönen 1988). Such three-phase systems have a minimum of three parallel wires carrying currents of the form

$$I_1 = I_{01} \cos (\omega t), \quad I_2 = I_{02} \cos (\omega t + 120°),$$

$$I_3 = I_{03} \cos (\omega t + 240°). \qquad (21)$$

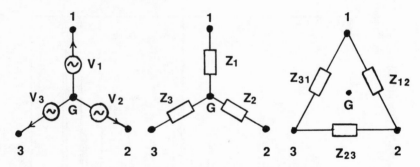

Figure 2.8. Generator (left) and load configurations (Zs) for the Y system (center) and the Δ (Delta) system (right). *Note:* G denotes ground wires.

That is, each current differs by a phase angle of 120°. In practice, power companies try to match the loads connected between these lines so that the three current amplitudes are nearly equal.

The two main systems in use (see figure 2.8) are referred to as the Δ (Delta) system and the Y system (or star system) for reasons that are self-evident from the geometric patterns in the figure. As illustrated in figure 2.8, the three-source electromotive-force (EMF) voltages are of equal magnitude with respect to the ground or neutral wire at the generator, but shift successively in phase by 120°. The Y system uses four wires in the transmission line. The center connection to the Y is a common return path (often labeled N for "neutral," or G for "ground"). In contrast, the neutral (or ground) connection is omitted in the Δ system, and only a three-wire transmission line is used. In each case, the connections from the generator labeled 1, 2, and 3 are attached through the transmission line to the corresponding terminals on the load impedance. In the Y system, the fourth wire connects the neutral point on the generator to the neutral point on the load. With regard to magnetic fields generated by the power line, there is no difference between the two systems as long as the three loads are balanced (that is, the three primary currents are equal).

The neutral wire on the transmission line in the Y system contains the sum of the three currents,

$$I_{01} \cos (\omega t) + I_{02} \cos (\omega t + 120°) + I_{03} \cos (\omega t + 240°), \tag{22}$$

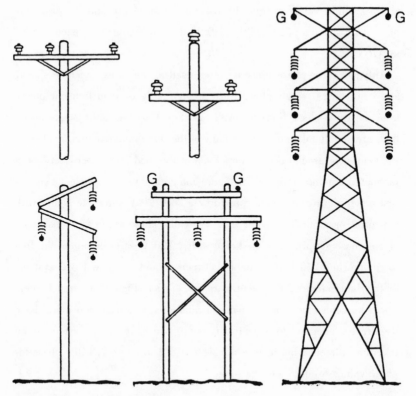

Figure 2.9. Representative support-pole configurations for three-phase Δ and Y systems *Note:* G denotes ground wires.

which, as may be seen from simple trigonometric identities, is zero when $I_{01} = I_{02} = I_{03}$. When the load is balanced, therefore, only three wires are carrying current in each system. The three current-carrying wires usually have about the same geometrical configuration in both systems, although in Y systems the ground (or neutral) wire is often mounted above the others for protection against lightning on long runs and mounted below the others on distribution poles in urban areas. In the latter case, the same neutral wire is often used for both the primary and the secondary side of the step-down transformers that provide 240/120-V lines to homes. (See the representative support-pole geometries for three-phase systems shown in figure 2.9.) Although considerable variation in pole height and wire spacing exists for 12-kV urban distribution lines, the typical pole height is about 35 ft.

(10 m), with a variation from 30 to 80 ft. (9 to 24 m), and the phase-to-phase wire spacing is about 36 in. (0.9 m), with a minimum spacing of 30 in. (76 cm).

Although three-phase lines are often used for factories, machine shops, and laboratories, single-phase lines are normally used to transmit power from the three-phase lines to dwellings. For Y systems, the single-phase lines are simply taken off between any of the three numbered terminals and the neutral or ground connection. Each of the load impedances between a numbered terminal in the middle figure and ground might represent a single-phase step-down transformer providing a 240/120-V supply to homes with a neutral or grounded center tap. Typically, three wires for a balanced single-phase supply enter the home with ± 120 V on each side of the line to ground, and 240 V is available between the outside pair. For Δ systems (at the right in figure 2.8), the primary side of such single-phase transformers would be connected between pairs of numbered terminals and would have no direct connection to a neutral (ground) wire. For either system, if we take the difference between terminals 1 and 2 in figure 2.8, the phase-to-phase voltage would be of the form

$$V_1 - V_2 = V_0 \cos (\omega t) - V_0 \cos (\omega t + 120°)$$
$$= \sqrt{3} V_0 \cos (\omega t - 30°), \qquad (23)$$

where V_0 is the voltage with respect to ground.

The 12-kV distribution lines commonly used in urban communities are Y systems, in which the rms phase-to-phase voltages typically vary from about 12.47 to 13.80 kV. Hence, the rms voltages with respect to ground are typically 7.20 to 7.97 kV. Although the transition from peak to rms currents and voltages simply involves dividing by a factor of $\sqrt{2}$, a more elaborate averaging is required to get rms values of the fields in three-phase systems. Because of the phase differences in eq. (21), the spatial distribution of the magnetic and electric fields varies spatially over the cycle.

The discussion so far has assumed that equal-load impedances are placed across the pairs of terminals on the three-phase line; that is, that the lines are balanced. This is the condition utility companies strive to achieve because it provides the least strain on the power generator and the lowest power loss in transmission. For the same transmitted power and line voltage, three-phase lines require less current per wire and generally result in lower magnetic (and electric) fields at ground level than single-phase line pairs.

Because a statistically significant difference in leukemia deaths was reported in Los Angeles between regions using Δ- and Y-system three-phase lines (London et al. 1991),[5] it is worth comparing stray magnetic field differences between the two systems when there is a load imbalance. The results from such a comparison depend critically on the wiring geometry in each case, something that is not adequately described in the article by London et al. For the sake of specific comparison, we shall consider the configurations described below. Because the magnetic fields vary spatially over the period of the line frequency, the comparison is presented as a table giving the rms values of the field magnitudes produced when 500-A rms flows through each primary line. The height of the middle wire is taken to be 10 m, and the wire separation to be 1 m. In the Y system, the fourth (neutral or ground) wire is assumed to be 2 m below the middle primary wire (see table 2.4).

In measuring such fields, it is important to use a meter that determines the resultant magnetic field from simultaneous measurement of the three separate orthogonal components. Variations on the order of 41 percent (a factor of $\sqrt{2}$) can arise from circular polarization effects if the determination is made from three separate measurements of the orthogonal field components using a single pickup coil. A number of three-coil magnetometers (the Dexsil Field Star 1000, or Magnum 310, for example) that permit computing the resultant rms magnitude with a microprocessor are available commercially.

Under balanced conditions, there is no difference between the magnetic-field distributions in the Y or Δ systems for the same spatial configuration. But there are differences when one line is overloaded. For purposes of comparison, we have assumed in table 2.4 that a balanced operation involves 500-A rms in each primary line and that an unbalanced operation has a 50 percent increase in the load on one single-phase line derived from the three-phase system. The primary wire is taken to be either at $x = -1$ m for the horizontal or at $y = 1$ m below the middle in the vertical configuration. The

Table 2.4. RMS Magnetic Fields near Ground for Balanced and Unbalanced Y and Δ Three-Phase Transmission Lines (mG)

						x (m)							Average (mG)
y (m)	−10	−8	−6	−4	−2	0	2	4	6	8	10	

Horizontal Configuration

Balanced Δ or Y

y (m)	−10	−8	−6	−4	−2	0	2	4	6	8	10	Average (mG)
3	12	15	20	26	32	34	32	26	20	15	12	22.3
2	10	13	17	21	25	27	25	21	17	13	11	18.3
1	10	12	15	18	20	21	20	18	15	12	10	15.3
0	9	10	13	15	16	17	16	15	13	10	9	13.0

Total average 17.2

50 percent unbalanced Δ

y (m)	−10	−8	−6	−4	−2	0	2	4	6	8	10	Average (mG)
3	15	20	26	34	41	43	40	33	25	18	14	28.0
2	13	17	22	27	32	33	31	27	21	17	13	23.0
1	12	15	19	22	25	26	25	22	18	15	12	19.3
0	11	13	16	19	21	21	21	18	16	13	11	16.3

Total average 21.7

50 percent unbalanced Y

y (m)	−10	−8	−6	−4	−2	0	2	4	6	8	10	Average (mG)
3	18	23	32	42	51	53	46	36	27	20	15	33.1
2	16	21	27	33	39	40	36	30	23	18	14	26.9
1	14	18	22	27	31	31	29	25	20	16	13	22.4
0	13	16	19	22	25	25	24	21	17	14	12	18.9

Total average 25.3

Vertical Configuration

Balanced Δ or Y

y												RMS
3	12	15	20	27	33	36	33	27	20	15	12	22.6
2	10	13	17	22	26	27	26	22	17	13	10	18.5
1	9	12	15	18	20	22	20	18	15	12	9	15.4
0	9	10	13	15	17	17	17	15	13	10	9	13.1

Total average 17.4

50 percent unbalanced Δ

y												RMS
3	15	19	26	35	43	47	43	35	26	19	15	29.3
2	13	17	22	28	33	35	33	28	22	17	13	23.9
1	12	15	19	23	26	28	26	23	19	15	12	19.9
0	11	13	16	19	22	22	22	19	16	13	11	16.8

Total average 22.5

50 percent unbalanced Y

y												RMS
3	8	11	14	18	21	22	21	18	14	11	8	15.3
2	8	10	12	15	17	18	17	15	12	10	8	12.7
1	7	9	10	12	14	14	14	12	10	9	7	10.7
0	6	8	9	10	11	12	11	10	9	8	6	9.1

Total average 11.9

Note: Middle wire is at $x = 0$; x is the horizontal distance to the middle wire; y is the height above the ground. RMS values of the field resultant are given.

Table 2.5. Maximum Magnetic Flux from Typical Wiring and around the Home

Wire Type	Wire Spacing (in.)	(cm)	Current (A rms)	RMS Magnetic Field (mG)	RMS Magnetic Field (mG)
Outside the Home				At 3 m	At 6 m
(vertical configuration)				(below wire)	
Pole-to-pole	$\approx$15	$\approx$38	500	42	10.6
Pole-to-home	$\approx$6	$\approx$15	100	3.4	0.8
Pole-to-home[a]	$\approx$1	$\approx$2.5	100	0.6	0.1
Inside the Home				At 0.5 m	At 1 m
Knob & tube[b]	$\approx$6	$\approx$15	20	24	6.1
Knob & tube[b]	$\approx$2.5	$\approx$6.3	20	10	2.5
Romex	$\approx$0.25	$\approx$0.6	20	1	0.3
BX cable[a]	$\approx$0.12	$\approx$0.3	20	<0.5	<0.1
Outlet strip	$\approx$0.5	$\approx$1.3	5	0.5	0.1
Lamp cord	$\approx$0.1	$\approx$0.2	1	<0.1	<0.1

[a]Twisted pairs; field represents upper limit
[b]Old wiring style; used until about 1930

unbalanced return is through the middle wire on the Δ system and through the fourth wire on the Y system. The differences between the overloaded Δ and Y systems are not large. It is easiest to see them by comparing the average values in table 2.4. Note that the variations between unbalanced Δ and Y systems go in opposite directions for horizontal and vertical configurations.

Magnetic Fields from Home Wiring

Several cases of interest involving parallel-wire lines arise in the distribution of electrical power in and around the home. Table 2.5 contains a summary of the common types of wiring and the maximum currents typically used. (Representative wiring configuration is reviewed in Croft and Summers [1992].) In some instances, further magnetic-field reduction arises from the use of twisted pairs of wires. Stray magnetic fields from wiring in contemporary homes are typically less than 1 mG. Houses with older "knob and tube" wiring often have fields $\approx$ 5 mG produced by wires going to ceiling

Table 2.6. Magnetic Fields under a Power Line arising from Parallel
Conduction through an Electrically Grounded Water Pipe

	RMS Magnetic Field (mG) at Specific Heights above the Ground			
R_{wire}/R_{pipe}	0 m	1 m	2 m	3 m
0.001	2	3	5	14
0.01	3	4	6	15
0.1	13	13	16	28

Note: The pipe is 8 ft. (2.44 m) below the ground and directly under a two-wire power
line carrying 100-A rms with a 6-in. (15 cm) spacing and located 4.5 m above the
ground. For No. 2 Cu wire, R_{wire} = 0.162 Ω/1,000 ft. (305 m); for No. 4 Cu wire,
R_{wire} = 0.259 Ω/1,000 ft. at 25°C (Baumeister and Marks 1967, 15-10).

and wall lighting fixtures. Fields from power lines in the street are typically
below 1 to 5 mG.

A common source of higher magnetic fields in power delivery to homes
arises from extra ground connections (or ground loops) through water pipes.
The magnitude of the effect depends on the relative resistance of the water
pipe (typically iron) versus that of the neutral wire (typically, No. 2 or No.
4 braided copper wire) in the power line. Table 2.6 has been computed for
several representative values. Clearly, rerouting or adding ground return
wires can produce background magnetic fields on the order of 10 mG in the
home.

Measured Magnetic Fields from Distribution Lines

Spot measurements of magnetic fields from 12-kV, 60-Hz, three-phase
distribution lines were made with a three-coil magnetometer in areas ranging
from Guilford, Connecticut, to Bryn Mawr, Pennsylvania. These locations
included New Haven, Bridgeport, New York City, and Philadelphia. The
fields at chest level were seldom in excess of about 7 mG and typically
ranged from 1 to 2 mG along urban streets. Fields of about 7 mG occurred
when two sets of distribution lines crossed at right angles and in rural areas
along long stretches of open road, where one distribution line with widely
spaced wires served an entire town. In most cases, the fields near ground

level from 12-kV urban-distribution lines were not much larger than typical background fields from house wiring (about 1 mG).

Fields from 240/120-V secondary lines from distribution-line transformers are often much higher. One pole containing a single transformer some 150 ft. (46 m) from the infamous power-transformer substation on Meadow Street in Guilford has two vertical conduits of five inches in diameter producing fields of about 160 to 180 mG at the pole, 15 mG at the curb, and 2.5 mG across the street, where the houses are located. At Yale University in New Haven, one similar underground line at an entrance to Dunham Laboratory produces about 10 mG at chest level, with resultant fields of about 6 mG in my own office. Similar lines in the ceilings of the main hallways of the engineering and computer science buildings at Yale produce fields in the 1 to 5 mG range.

Apart from electrified railroads, magnetic fields of larger than 10 mG were seldom encountered in normal travel except under high-voltage transmission lines. These lines occasionally cross major highways, with resultant fields ranging from about 15 to 90 mG. Lines running parallel to major highways in New Jersey typically produced fields on the order of 25 mG inside a small Subaru station wagon.

Magnetic Fields from Current Loops

The magnetic field from a current loop can best be evaluated directly from the differential form of the experimental law discovered by Biot and Savart:

$$d\mathbf{H} = I\,\frac{d\mathbf{L} \times \mathbf{r}}{4\pi r^3} \quad \text{[SI units]},\qquad (24)[6]$$

where I is the current flowing in a wire with element of length and direction given by the vector $d\mathbf{L}$, and $\mathbf{r}$ is a vector from the current element to the point of observation. (The 4π factor is unit-dependent and is present in SI units, where $\mathbf{H}$ is in ampere-turns per meter [At/m] and $\mathbf{I}$ is in amperes.) From eq. (24), the magnetic induction field at the center of a circular current loop of radius a is

$$B_z = \mu_0 I/2a \quad \text{[SI units]},\qquad (25)$$

where I is the current in amperes and a is in m. From symmetry the field is along the z direction, which is taken to be perpendicular to the plane of the coil. Evaluating the cross product at some general point z along the axis of a circular coil yields

$$B_z(z) = \frac{\mu_0 I}{2} \frac{a^2}{(a^2 + z^2)^{3/2}} .\tag{26}$$

If N closely spaced turns of wire are applied on the same circle of radius a, the magnitude of the field is multiplied by N.

Unfortunately, it is not possible to obtain an equivalent closed-form expression for the field at a general point away from the axis of symmetry. At large distances — $r^2 \gg a^2$ — the coil looks like a magnetic dipole, and the magnetic induction field has components

$$B_z = M\,(2z^2 - x^2 - y^2)/r^5,\; B_x = M\,3xz/r^5,\; B_y = M\,3yz/r^5,\tag{27}$$

where x and y are the transverse coordinates in the plane of the coil, $r^2 = x^2 + y^2 + z^2$, and $M = \mu_0 I a^2/4$ is the magnetic-dipole moment from the current loop. Adding several coplanar and coaxial coils of different radii together increases the size of the magnetic-dipole moment M; this increase goes up linearly with the number of coils in the limit that $z^2 \gg a^2$. It is thus particularly easy to estimate the magnetic field from a nest of coaxial coils at large distances compared with the coil radii. The magnetic-dipole approximation is not useful, however, for determining the field in close proximity to a coil.

Even for a circular current loop, the evaluation of the near magnetic field off the axis of symmetry involves something equivalent to infinite series approximations or the numerical evaluation of elliptic integrals.[7] It is easier and more general to use a computer to evaluate the integral of the Biot-Savart relation in eq. (24) over the contour of the coil directly. The latter approach has been used here to compute the field from several coil geometries.

The magnetic-field lines for a circular coil are shown in figure 2.10. The results are rotationally symmetric about the vertical axis and are also symmetric about the median plane (indicated by the horizontal line in the

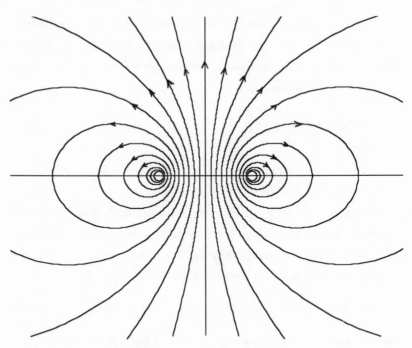

Figure 2.10. Magnetic-field lines for a circular coil. *Note:* Coil cross section is indicated by the two circles on the horizontal line through the median plane. The current flows out of the diagram at the left and flows in at the right. The coil axis (*z*-axis) is vertical.

figure). The coil cross section is indicated by the two circles on the median-plane axis. It is assumed that the current flows out of the diagram at the left and back into the diagram at the right. As is apparent from the crowding of the flux lines, the field is again strongest near the wires, and its strongest components are in the vertical direction. Numerical values of the field are given in table 2.7 for the specific coil configurations and currents.

Magnetic Fields from Electric Hot Plates

The magnetic-field amplitudes for the three coil configurations shown in figure 2.11 and summarized in table 2.7 have been computed as a function of position for a current of 10-A rms flowing through the coil in each case and for an outer radius of about 10 cm.

Table 2.7. RMS Magnetic Fields as a Function of Position for the Three Planar-Coil Configurations of Figure 2.11 (mG)

	Vertical Axis, z (cm)										
Configuration	0	10	20	30	40	50	60	70	80	90	100
Circular coil	628	222	56	20	9	4.7	2.8	1.8	1.2	0.8	0.6
Spiral	5,435	1,039	227	78	35	18	10.8	6.9	4.7	3.3	2.4
Bifilar spiral	713	87	27	10	4.7	2.5	1.5	0.9	0.6	0.4	0.3

	Horizontal Axis, x (cm)										
	0	10*	20	30	40	50	60	70	80	90	100
Circular coil	628	750	54	13	5.3	2.6	1.5	0.9	0.6	0.4	0.3
Spiral	5,435	11,800	221	58	25	13	7.8	5.1	3.5	2.6	2.0
Bifilar spiral	713	6,130	35	7.7	3.0	1.5	0.8	0.5	0.34	0.23	0.2

Note: The values (*) at 10 cm on the horizontal axis fall near the outer radii in each case and are not well defined.

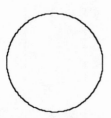 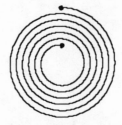 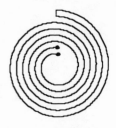

Figure 2.11. Three coil configurations: circle (left), spiral (center), and bifilar spiral (right)

The spiral coil resembles the heating elements found in many electric stoves and hot plates. At 10 A on a 120-V line, it would produce about 1.2 kW. Similar configurations are used at 10 A at 240 V to provide heating elements of about 2.4 kW. As shown later, in table 2.8, the magnetic fields from the spiral hot plate are among the largest encountered in the normal home environment. The field in this case no longer has perfect axial symmetry, and small differences in the field-line patterns arise for different choices of the horizontal axis.

AC Magnetic-Field Reduction

Although there is no clear reason to assume that fields like those from a spiral hot plate represent a health hazard, they could easily be reduced by close to an order of magnitude by substitution of a bifilar-wound spiral. The bifilar-wound spiral occupies roughly the same space and would produce about the same heat as the spiral element used in most electric stoves.

Magnetic-field reduction at 60-Hz by a factor of from two to three can easily be obtained with many devices by wrapping them in a layer of "mu metal" — a thin, light-weight, high-permeability material that serves to trap magnetic flux. Since mu metal is not transparent this method cannot be used on the viewing screen of a television set or a video display terminal. More substantial reductions at 60 Hz through this approach require fairly heavy ferrous casings. The simplest method of field reduction in many cases is through use of twisted-pair or bifilar wiring, which avoids large loops in the wiring. Bifilar wiring has already been adopted by the electric-blanket industry as a precautionary measure. As is well known to designers of audio equipment, the simple avoidance of "ground loops" is one of the most cost-effective ways to minimize stray magnetic-field pickup and generation. (A ground loop is an unintended conduction path between electrical grounds that gives rise to an extra current loop, which might either generate external magnetic fields or provide unusual sensitivity to such fields by electrical pickup through the Faraday effect.)

There are many reasons other than concern over health for minimizing unnecessary 60-Hz magnetic-field pollution. To take an example, such fields are the primary cause of "hum" at the line frequency in audio equipment. They also limit the ability to perform spectral analysis or digital signal processing with computer-based equipment in the audio range. For example, analog-to-digital converters with 16-bit resolution should ideally provide about a 99-dB rms signal-to-noise ratio at maximum level.[8] When placed inside a personal computer, 60-Hz pickup in such devices is often down by only about 40 dB from maximum signal level. Similar pickup on telephone lines was one of the reasons the Metro-North commuter railroad adopted a new wiring geometry to reduce magnetic fields when they converted from 25 to 60 Hz.

Fields from Household Appliances and Office Equipment

Magnetic fields comparable to those of the electric hot plate are to be expected from any electrical heating device comprised of coils of wire. In general, heating appliances that contain coils with a large number of turns produce proportionately larger magnetic fields. These fields are largest near the object, but they eventually fall off according to the $1/r^3$ law predicted by the magnetic-dipole approximation. Again, the use of either bifilar windings or pairs of coils with opposing current loops (which could easily be introduced in the manufacture of electric broilers and toaster ovens) can typically reduce the size of these fields by an order of magnitude.

A three-coil magnetometer (the Field Star 1000) was used to make spot measurements of the magnetic fields produced by common 60-Hz household appliances and office equipment. The values were checked and the third harmonic measured using a different three-coil magnetometer with separate filters at 60 and 180 Hz (the Dexsil Magnum 310). The principal errors in measurement arose from spatial resolution at distances of less than 5 cm because of the size of the meter coils and because of background fields of 0.4 to 1 mG from house and office wiring, which were subtracted from the measured values for the appliances (see table 2.8). The spatial resolution of the meters was inadequate to determine the rapidly increasing field at small distances from the spiral hot plate. (Compare with the results calculated in table 2.7.) With the exception of the video display terminals and the television set (which were measured from the center of the screen), the location of the maximum field was first determined, and the fields were then measured outward from that point. The fluorescent light was a ceiling fixture containing four 40-W bulbs, and the fields were produced for the most part by the transformer. Fields of smaller than 1 mG are not listed, except for the telephone and a liquid-crystal display from a portable computer. In some instances, magnetic fields encountered at 180 Hz (the third harmonic of the power-line frequency) were down by factors ranging from about two (an FM wireless intercom) to ten (a microwave oven) from the peak fields at 60 Hz. These were primarily instances in which nonlinear load resistances (discharge tubes, rectifiers, and the like) were involved in the appliance. The airport metal-detector field was at 180 Hz. The fields from video display terminals are extremely complex and will be discussed in a later section. Variations of $\approx$ 10 percent were encountered in repeated measurements for some of these devices. The "60-Hz" fields from the DC electric razor were surprising. The battery-powered motor ran at a frequency close 30 Hz, and the actual magnetic fields at 30, 60, and 180 Hz found at 2 cm from the razor were about 32, 2, and 2.5 mG, respectively, as determined by measurements made with three different filters. The largest fields encountered were from the bulk magnetic tape eraser, a unit designed to erase ¼-in.

tape on open reels. (From the data in table 2.8, the electric can opener would probably make a reasonable substitute.)

Magnetic Fields from Power Transformers and Motors

Large coils are contained in the pole-mounted power transformers that convert the 12 kV distribution lines to the 240/120-V lines that serve private homes. Although the many turns used in these transformers result in high internal magnetic fields, the coils are wound on high-permeability laminated cores that prevent significant escape of magnetic flux outside the transformer.[9] The cores are contained within steel cases. Stray ground-level magnetic fields produced by these transformers are therefore likely to be no larger than those produced by the power lines themselves. Note, for example, that spot measurements made over a year's time with Field Star 1000 and Magnum 310 magnetometers at the power substation on Meadow Street showed that the maximum fields next to the fence around the transformers amounted to only about 17 to 25 mG at 60 Hz ($\approx$ 3 mG at 180 Hz) and fell off to about 5.5 to 6.5 mG at the edge of the road. The field at the curb on the other side of the street — where the only houses are located — ranged from about 1.5 to 2.5 mG along the entire block.

These transformers can get fairly hot during normal operation, however, and some of the cooling oil contained in the transformer casing can evaporate when they overheat. Because some transformer oils used to contain polychlorobenzene (PCB) and other chemicals that could be carcinogenic (a controversial speculation in itself), cancer could be induced by an entirely nonelectromagnetic process in people living near power-distribution points. Under normal operation, the evaporation of oil should be minimal because the cover plates on the transformer casings are sealed with gaskets. But many transformers are fitted with pressure-release valves that permit oil to escape during overheating. In 1979 the Environmental Protection Agency recommended regulations to limit the use of PCB-contaminated oils in transformers, defining three categories of transformer: "non-PCB" (less than 50 ppm); "PCB-contaminated" (50–500 ppm); and "PCB" (greater than 500

Table 2.8. Maximum RMS Magnetic Fields as a Function of Distance from Ordinary Household Appliances and Office Equipment

Appliance	RMS Magnetic Fields (mG) at Specific Distances (cm) from the Appliance					
	2	4	10	20	40	100
Bulk magnetic tape eraser (top)	19,900	16,400	10,800	3,800	580	100
Electric can opener (back)[a]	12,400	10,400	2,080	352	52	2
DC toothbrush charger (side)[a]	6,000	4,000	400	71	9	
Spiral hot plate (top)	1,570	1,160	750	31	4	
Kitchen blender[a]	1,320	1,000	180	34	14	
Microwave oven (rear side)[a]	1,050	820	350	140	28	5
FM wireless intercom (top)[a,d]	700	300	38	8	1	
Clock radio (top)[a,d]	248	188	22	5	1	
Coffee maker (side)	230	90	26	7	1	
Electronic bug zapper[a]	160	104	20	7	1	
Electric blanket, 180 W, queen size	140	70	13	6	2	
View-Graph projector (top)[h]	116	112	40	14	5	
Electric blanket, 135 W, twin size	104	55	4	3	1	
DC electric-shaver charger (side)[a]	86	34	5			
Furnace fan (front)	76	36	11	5	1	
Electronic typewriter (top)	70	56	30	14	6	2
Office copier, running (top)	61	39	34	19	5	
Clock radio (side)[d]	48	11	4	1		

Device					
Attic exhaust fan (front)	45	21	6	2	
Fluorescent ceiling light, 160 W (below)	40	34	26	18	6
Circuit breaker, 100 A (front)	40	31	21	12	2
Stereo amplifier, 200 W (front)[a]	39	27	13	5	1
Office copier, idling (top)	35	30	18	10	3
VDT (video display terminal), 13-in. color, 1981 (front)[a,f]	34	27	7	1	
VDT portrait monitor (front)[a,f,g]	32	24	11	6	2
Toaster oven (front)	31	26	14	7	1
Airport metal detector (180 Hz)[b]	30	26	23	11	1
TV, 32-in. color, 1990 (front)[a,f]	29	23	15	11	4
Washing machine (bottom, side)	30	22	11	5	2
VDT 13-in. color, 1988 (front)[a,f,g]	24	17	10	5	1
Electric typewriter (rear, side)	21	13	9	4	2
Heating pad (top)	20	11	3		
VDT, 11-in. B&W, 1981 (front)[a,f]	16	10	7	2	
Digital-audio-tape player (front)	15	10	3		
Hair dryer, 1500 W (side)[a]	15	8	3		
Refrigerator (rear, side)	13	11	8	5	
VDT, 12-in. color, 1991 (front)[a,f]	12	10	5	1	
VDT, 13-in. color, 1992 (front)[a,f]	12	10	4	1	
Laser printer (side)[a,e]	12	4	3	2	
Compact-disc player (front)	10	6	3	1	

(continued)

Table 2.8. Continued

Appliance	RMS Magnetic Fields (mG) at Specific Distances (cm) from the Appliance					
	2	4	10	20	40	100
Four-slice toaster (side)	8	4	1			
Audio-cassette player (front)	6	5	3	1		
VDT, 11-in. yellow, 1991 (front)[a,f]	6	4	2			
VDT, 13-in. color, 1993 (front)[a,f,g]	5	4	2			
Light bulb, 100 W (side)	4	2.5	1.5			
Electric mattress cover (top)	4	2				
FAX machine (top)[a,e]	3	2	1			
DC electric shaver, running (top)[c]	2	1				
Color scanner (side)	1					
Telephone	0.1					
Liquid crystal display, 10-in. (top)	0.1					

Note: The fields are arranged in order of decreasing value at the source. They were measured with three-coil magnetometers equipped with 60- and 180-Hz filters.

Except where noted, fields were at 60 Hz.

[a]Appreciable third harmonic (180 Hz)

[b]Main field at 180 Hz

[c]Main field at 30 Hz

[d]Device off but plugged into line voltage; negligible difference when turned on

[e]While printing; less than 1 mG while idle but turned on

[f]Video and TV displays by different manufacturers. The field in the screen has a strong horizontal component corresponding to the vertical deflection field.

[g]70-Hz vertical-sweep frequency

[h]Primarily from cooling fan

ppm). The difficulties of identifying and testing such transformers are formidable. When the Utah Power and Light Company instituted a program to
remove PCB transformers by the end of 1988, about 198,000 transformers
throughout the state had to be tested (Mills and Rhoads 1985). Implementing
this kind of program depends on finding economical procedures for moni-
toring low PCB levels, and the design of tests that will be reliable at the
50-ppm level is a difficult undertaking.[10] But it should be emphasized that
cancers that might be produced by possible chemical carcinogens associated
with the presence of power transformers (or power lines) cannot be blamed
on the electromagnetic fields themselves.

Large coils in motors are also wound on high-permeability laminated
cores, which prevent the significant escape of magnetic flux outside the
motor housing. Even without the high-permeability core, the magnetic fields
from both motors and transformers would decrease with distance r as $1/r^3$:
that is, the magnitude would be given by the field from a magnetic dipole
in eq. (27), where the parameter a is the coil radius and $r \gg a$. (See the
data for the view-graph projector, the furnace fan, the attic exhaust fan,
and the washing machine in table 2.8.)

Magnetic Fields from Television Sets and Video Display Terminals

Magnetic fields from television sets and video display terminals are extremely complex. There are several basic sources of these fields, with
frequencies that spread over both of the bands in figure 1.1. In addition to
fields from the power source at 60 Hz, time-dependent magnetic fields are
associated with the scanning process used to generate the visual display.
Most current television sets and video display terminals contain cathode ray
tubes that employ magnetic deflection of the electron beam. Sawtooth-
shaped, time-dependent magnetic fields are used to accomplish the scanning
in both the horizontal and vertical directions, and these waveforms are rich
in harmonic content. The method of horizontal interlace scanning currently
used on television sets in the United States to provide a basic frame-
repetition frequency of 30 Hz requires sawtooth waveforms with a funda-

mental frequency of 60 Hz associated with the vertical-scanning process. Although the deflection coils are at the back of the tube, fringing fields from these coils are encountered at the screen. As a result, 60-Hz magnetic fields are encountered at the front of the set from both the power-line source and the vertical scanning process. This complexity introduces substantial opportunity for error in magnetic-field measurement. In addition, many video display terminals use progressive (noninterlace) scanning at a repetition rate of about 70 Hz to minimize flickering sensations when the terminal is viewed. Those displays include harmonics of both 60 Hz from the power supply and 70 Hz from the vertical scanning process. An unusually good filter is required to separate those components, and commercially available magnetometers like the Magnum 310 do not have adequate resolution. Further, circular or elliptic polarization can result from the superposition of the different 60-Hz sources. Consequently, it is desirable to measure fields with a three-coil magnetometer equipped with several different filters so that the true resultant field can be found from the separate spatial components at each frequency. In principle, one could also use the polarization of the field to help identify different sources. At the center of the screen, the low-frequency vertical scanning field from the deflection coils should be linearly polarized in the horizontal direction.

Representative measurements of nominal "60-Hz" magnetic fields are contained in table 2.8 for a variety of video display terminals made by different manufacturers over the past ten years, as well as for a contemporary thirty-two-inch color television set. These all contained significant amounts of third harmonic and were partially polarized in the horizontal direction. The values measured at 60 Hz ranged from about 6 to 34 mG at the screen to about 0.5 to 11 mG at 40 cm. The only significant trend in the data was that fields from the most recently made video display terminals tended to be the lowest. (The manufacturers are clearly reacting to public concern and in some cases specifically advertise "low emissions.")

The difficulty in making precise measurements of the field is illustrated by results for a thirteen-inch color video display terminal using 70-Hz vertical scanning. Measurements made with a Magnum 310 provided with

Table 2.9. Field Components Measured at the Screen from a 13-inch Color
Video Display Terminal

Filter	RMS Magnetic Fields (mG) at Specific Orientations			
	Vertical	Horizontal	Out of Screen	Resultant
Broadband	10	40	22.4	46.9
60 Hz	3.7	20	12	23.6
180 Hz	0.16	6.6	3.7	7.6

Note: The measurements were taken with a Dexsil Magnum 310 magnetometer. The video
display terminal uses a 70-Hz vertical-scan frequency and is powered at 60 Hz. These
measured values probably arise from the following actual resultant field magnitudes:
12.3 mG at 60 Hz and 2 mG at 180 Hz from the line frequency; 14.3 mG at 70 Hz, 7.2 mG
at 140 Hz, 4.8 mG at 210 Hz, 3.6 mG at 280 Hz, and so on, down to 0.1 mG at 700 Hz
from the sawtooth for vertical scanning.

three different filters are shown in table 2.9. Direct measurements of the
field magnitude at the center of the screen gave values of about 47 mG with
the broadband filter, 24 mG with the narrow-band 60-Hz filter, and 7.6 mG
with the narrow-band 180-Hz filter. None of these filters can resolve fields
at the various possible harmonics of 60 and 70 Hz, and the measured values
are not very meaningful by themselves. But we can use the known filter-
response functions to unfold the data. Although we cannot determine more
than three unknown quantities from only three measurements, we expect in
advance that the field magnitudes at harmonics of 70 Hz will correspond
approximately to the coefficients of the Fourier series for a sawtooth and
vary as $1/n$ where n is the harmonic number.[11] Hence, the responses of the
broadband 60-Hz filters may be used to determine both the 60-Hz field and
the common coefficient for, say, the first ten harmonics of the sawtooth at
70 Hz. In principle, we should determine each of the spatial rms components
separately and then take the square root of the sum of their squares at each
frequency.[12] In the present case, however, the degree of polarization is
roughly the same for the major components of each filter output, and
additional refinement is unnecessary.

Using the total rms magnitudes from each filter output for illustration,
we arrive at two equations with two unknowns of the form

$0.8378\,A + 2.2855\,B = 46.9$ mG (broadband filter)

and

$A + 0.7879\,B = 23.6$ mG (60-Hz filter),

where A is the unknown field magnitude at 60 Hz, B, for the fundamental frequency of the 70 Hz sawtooth. The numerical coefficients are determined by adding the contributions of the two filter-response curves for the first ten harmonics of the sawtooth. Solving the two equations simultaneously yields $A = 12.3$ mG for the 60 Hz field and $B = 14.3$ mG for the leading term in the sawtooth. We then compute the response of the 180-Hz filter to the assumed form of these fields, using the values of A and B just determined. This gives us a third equation,

$0.0123\,A + 0.3848\,B + C = 7.6$ mG (for the 180-Hz filter),

where the numerical coefficients are based on the 180-Hz response curve and where C is the unknown value for the third harmonic of the 60-Hz fields from the power source. Combining these results, we find $C \approx 2$ mG, which gives us a reasonably good picture of the magnetic-field variation with frequency from about 60 to 700 Hz in steps of 70 Hz. Of course, we could easily work out the requisite coefficients from the filter curves provided by the manufacturer for sawtooths at any other fundamental frequency in the range covered by the meter. (A more reliable, but much more expensive way to come up with equivalent data is to attach the test coils to a spectrum analyzer.)

As noted, a strong linear polarization in the horizontal direction at the center of the screen arises from fringing fields produced by the vertical scanning coils. This effect is illustrated in table 2.9 for the example just discussed.

The values for 60-Hz fields of video display terminals in table 2.8 are significantly lower than other published values. Rosch (1990), for example, reports values of about 5 to 128 mG at 24 in. (61 cm) from the screen for different monitors, whereas Schnorr et al. (1991) report fields of about 5 to

50 mG at the same distance. These authors provide little information on their method of measurement, however, and it is unclear what corrections have been made for frequency response and spectral content.

All these authors agree that the strongest 60-Hz fields occur near the rear and sides of the monitor. Shielding by mu metal, which is relatively impervious to magnetic fields, can reduce the fields by factors of close to two or three. Of course, this technique cannot be applied on the screen, where limiting values will be determined by the fringing fields from the deflection coils. By contrast, computer monitors with liquid-crystal displays produce fields substantially smaller than 1 mG.

A potentially more serious source of magnetic fields exists in connection with the sawtooth voltage waveform used to produce *horizontal* scanning in the monitor. Here the frequencies are much higher and the electric fields that would be generated in the body because of the Faraday effect are proportional to the frequency through the rate of change of the field.

Because a rapid "flyback" pulse occurs at the end of the horizontal-scanning sawtooth in 5–11 μs, a sharp transient is generated periodically at the horizontal sweep frequency in the flyback transformer. This transformer could be a strong source of magnetic fields over a broad frequency spectrum. Yet high frequency fields produced by this transformer tend to be damped out by eddy currents generated in surrounding metal components of the chassis. To take a worst-case limit, suppose that peak fields as large as those reported by Schnorr et al. (1991) — namely, 50 mG at the operator's location — were actually present at the horizontal sweep frequency (15.75 kHz). The peak electric field produced through the Faraday effect by a flux of 50 mG at 15.75 kHz over a 10-cm circle would be about 0.025 V/m, or about 0.1 E_{kT} in the same bandwidth from thermal noise. (See the discussion of eq. [72] below.) So even in the worst case, the electric fields generated would be inconsequential in comparison to unavoidable natural sources. In addition, the fields at the horizontal sweep frequency at the viewer's position are likely to be much smaller than in this particular estimate.

One can estimate the limiting size of the fringing field in the screen direction using basic laws of physics. The deflection coils are located in something known as the deflection yoke at the neck of the cathode ray tube and are passed by the electron beam after it has left the electron gun and entered the final anode enclosure. The electrons at that point have reached their maximum kinetic energy. The coils are typically located about 20 cm (in video display terminals) to about 35 cm (in large color television sets) from the viewing screen and are folded around the neck of the cathode ray tube; the coil radii are about 2.5 cm. The coil axes are perpendicular to the electron beam in order to provide horizontal and vertical deflection. The magnetic-field change may be estimated from the magnetic-dipole approximation discussed in eq. (27), by use of the magnetic force in eq. (8) to determine the magnetic field required to produce the beam deflection. The beam-deflection angle in most contemporary television sets is about $57°$ (Bridgewater and Fink 1974). Sharper focusing, brighter images, and larger screen sizes have pushed the accelerating potentials up to the 35- to 50- kV range.

As a worst-case upper limit, consider a 50-kV electron beam. From the relativity theory, we know that the mass of the electron has gone up by about 10 percent ($m/m_0 \approx 1.1$) and that the electron velocity is about $0.41c$. For a deflection coil diameter of $2a \approx 5$ cm, the electrons go through the coil region in a time Δt of about 0.4 ns, and the force of the magnetic field must bend them through $\pm 57°$ in that time. The peak magnetic field required is given approximately by

$$\Delta v_x/v \approx \sin 57° = qB \, \Delta t/m \tag{28}$$

and amounts to $B \approx 13.1$ milliteslas (mT) $= 131$ G produced near the back of the monitor. The magnetic-dipole moment of the coil is $M = B_z(0)$ $a^3/2 \approx 1$ kG-cm^3. (See eqs. [25] and [27] and related discussion.)

For horizontal deflection on the television screen, the z-axis of the deflection coil (see fig. 2.10) is in the vertical direction, and the main component of the field in the plane of the screen is in the opposite vertical direction in front of the set, with an approximate magnitude given by

$B \approx M/R^3$, where R is the distance to the center of the deflection coil. (Note that the fields associated with both vertical and horizontal deflection should be of about the same magnitude.) Directly in front of the screen, $R \approx 30$ cm and $B \approx 5$ mG. Forty centimeters away from the screen, $B \approx 0.4$ mG, and the field rapidly drops off to even more negligible values. This worst-case limit gives fields that are actually smaller, but of comparable magnitude to, the measured "60-Hz" values in table 2.8.

The frequency spectrum of the field is complex because sawtooth waveforms are used to produce both the horizontal and vertical scanning. The horizontal-sweep frequency is higher than the vertical-sweep frequency by a factor of about 262, and each sawtooth waveform is rich in harmonics. An ideal periodic sawtooth waveform consists of a linear ramp voltage with a very short flyback time, and it has a spectral power density which approaches $1/n^2$ at the frequency nf_0, where f_0 is the repetition frequency of the sawtooth, and $n = 1, 2, 3, \ldots$ is the harmonic number.[13]

Television sets used in the United States produce 525 horizontal lines at 30 frames per second; thus, the fundamental frequency of the sawtooth is $f_0 \approx 15.75$ kHz. An ideal sawtooth at this fundamental frequency would have a spectral power density that decreases by only 30 dB at 0.5 MHz. As noted, 5–11 μs of the 63.5-μs horizontal-scanning sawtooth period are consumed by the flyback time, during which the horizontal synchronization and color-burst pulses occur.[14] As shown by the histogram in figure 2.12, which was computed for a 5-μs flyback time, the actual spectrum falls off more rapidly with increasing frequency than does the spectrum for the ideal sawtooth. (The spectrum of the flyback transformer pulse would extend to much higher frequencies.) Although the magnetic fields generated by the vertical-deflection coils are of about the same magnitude, the spectral components from their fields are scaled down by a frequency ratio of 262.5 from the values in figure 2.12.

Instead of the 525 horizontal lines and 30-Hz frame rate of television sets in the United States, those in the United Kingdom use between 405 and 626 horizontal lines at a 25-Hz frame rate, while those in South America and Japan use 525 lines at a frame rate of 25 Hz and in France and French

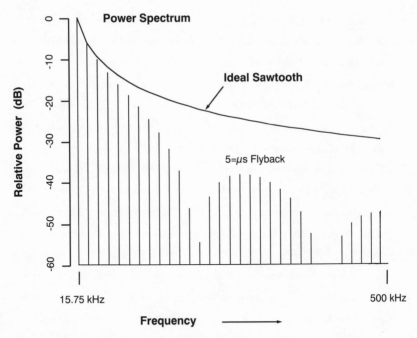

Figure 2.12. Power spectrum for ideal sawtooth (solid line) and for horizontal sweep with 5-μs flyback time (histogram)

dependencies use from 625 to 819 lines at a 25-Hz frame rate. In Japan engineers are also currently experimenting with a satellite-based, high-definition television that would produce 1,125 horizontal lines. In all these cases, the horizontal sweep frequency is equal to the number of horizontal lines multiplied by the frame rate. The vertical sweep frequency for interlace scanning systems is twice the frame rate. Hence, the frequency spectrum from vertical scanning falls in the low-frequency band discussed in this book, and the frequency spectrum from horizontal scanning falls in the higher-frequency band (see fig. 1.1). The standard for high-definition television recently adopted by the U.S. electronics industry incorporates progressive (noninterlace) scanning at a 60-Hz refresh rate.[15] There, fields from vertical scanning will have the same (60-Hz) fundamental frequency, while those for horizontal scanning will be shifted up to about 66 kHz. Current high-resolution video display terminals have between 480 and 1,040 lines, some with noninterlace vertical scanning frequencies above 70 Hz. The horizontal scanning in some currently available computer displays thus already encompasses fundamental sawtooth frequencies up to about 62 kHz.

Electric fields induced by the Faraday effect depend on the time-derivative of the magnetic field and have a relative power spectrum shifted toward higher values at high frequencies. (See the spectrum for the waveform used in therapy for fractured bones illustrated in figure 1.3.) As noted above, the electric fields from the fundamental horizontal sweep frequency would just become comparable to unavoidable internal thermal fields at a level of about 500 mG.

In a comparison study involving 882 pregnant telephone operators, half of whom used video display terminals, Schnorr et al. (1991) concluded that routine exposure to electric and magnetic fields from the terminals was not associated with an increased risk of spontaneous abortion. Studies conducted by Stuchly et al. (1988) in which rats were deliberately subjected for many days to 17-kHz sawtooth magnetic fields with peak intensities of 660 mG (above the thermal-field limit but far greater than those measured from video display terminals) showed no discernible effect produced by the magnetic fields.

Electric Fields

The analysis of electric-field lines is treated in detail in classical treatises on electromagnetic theory, like Jeans (1915). As was done to make the illustrations of magnetic-field lines shown above, this process may be greatly simplified by programming a computer to plot lines numerically through a series of successive steps taken along unit vectors in the direction of the field (Bennett 1976, 271). Of course, even with a computer, an analytic solution for the field expedites the procedure. Again, as with magnetic-field problems, there are several equivalent ways to calculate the electric fields from wires. The simplest approach for determining fields from straight power lines appears to be a combination of two of these methods: the "method of images," in which the electric fields are calculated from a set of linecharge distributions arranged to provide a zero-potential ground plane; and scalar potential theory, in which the capacitances necessary to evaluate the charges needed in the method of images are computed.

Electric Fields from Straight Wires

At the most direct level, the electric field from a charge distribution may be calculated from the second of Maxwell's equations in table 1.2. This equation contains Coulomb's law and is most usefully applied through Gauss's theorem when the charge distribution is known. When applied to an infinitely long, isolated straight wire in free space, the electric field is given by

$$E_r = q_L/2\pi\epsilon_0 r \quad [\text{V/m}], \tag{29}$$

where q_L is the constant charge per unit length along the wire. From symmetry, we know that the electric field E_r is in the radial direction and falls off as $1/r$, where r is the distance from the axis of the wire. Its magnitude can be found in specific cases from the capacitance C_L per unit length of the line with respect to ground, where $q_L = C_L V$ and V is the potential of the line with respect to ground.

There is no isolated single-line charge in the present problem, however. In general, at least two line charges are required to include the perturbing effects of a conducting ground plane (that is, the earth) below the wire. The result in eq. (29) is primarily useful when applied through the method of images (Maxwell 1873, chap. 11).[16] If we were to imagine a negative (mirror image) line charge $-q_L$ placed an equal distance below ground, for example, the sum of the two separate fields would produce a planar, zero-potential surface halfway between the two line charges. The vector superposition of those two fields (which are each directed radially from their respective line-charge axes) yields a solution above the ground that is formally equivalent to the solution for a single line charge above the conducting plane. By the superposition of the electric fields from successive pairs of such image-line charges, we can determine the total electric field for any number of parallel straight wires. The mechanism for doing this is similar to the one used earlier in this chapter to determine the vector superposition of magnetic fields. The total electric field from a group of n parallel line charges is given at the point $\mathbf{P}$ by

$$E = \sum_{i=1}^{n} Q_i \mathbf{r}_i / r_i^3, \tag{30}$$

where $\mathbf{r}_i = \mathbf{P} - \mathbf{W}_i$ and the vector $\mathbf{W}_i$ points to the location of the ith wire or line-image charge, and $\mathbf{P}$ is a vector defining a particular point in space from a common origin in a plane perpendicular to the wires. The term

$$Q_i = q_L / 2\pi\epsilon_0 = C_i V_i / 2\pi\epsilon_0 \tag{31}$$

contains the line-charge density per unit length and the other constants in eq. (29). C_i and V_i are the capacitance per unit length and the voltage with respect to ground for the ith wire. The potential on any conducting surface is constant and is determined by the applied voltage.

Potential theory provides a useful auxiliary method for treating these problems. It follows from Maxwell's equations that the electric field $\mathbf{E}$ in a charge-free region between conductor surfaces may be determined from a scalar potential ϕ through the relations

$$\mathbf{E} = -\nabla\phi, \quad \text{where } \nabla^2\phi = 0. \tag{32}$$

The solution of a particular problem thus can be obtained by solving Laplace's equation (the second relation in eq. [32]) for a potential ϕ which reduces to specified constant values on the surfaces of the conductors. These solutions are unique for a given set of potentials, and closed-form solutions to this type of boundary-value problem have been found for many geometric configurations. This approach works best when the conductor shapes are symmetric in the coordinate system used to analyze the problem. In those instances, known solutions to Laplace's equation may be employed directly to obtain the potentials, and hence the electric fields.[17] In several of the two-dimensional problems discussed here, the capacitances needed in the method of images may be determined in closed form from complex conjugate potential theory (Maxwell 1873, chap. 12).[18]

Electric Fields from a Single Wire above the Earth

Solutions of Laplace's equation in cylindrical coordinates show that the capacitance per unit length of a single wire at a height h above the ground is given by

$$C_L = \frac{2\pi\epsilon_0}{\log_e(2h/a)} \quad \text{[farad/meter, F/m]}, \tag{33}$$

where a is the radius of the wire, and the approximation is valid in the limit that $a^2 \ll h^2$ (Harnwell 1938, 37–44; Page and Adams 1945, 90, 106–113). If, for example, we take $a = 1$ cm, the capacitance is $C_L = 7.3$ picofarads per meter (pF/m) at a height $h = 10$ m. When the overhead line is charged to a voltage V_0 with respect to the ground plane, the coefficient in eq. (30) becomes

$$Q = q_L/2\pi\epsilon_0 \approx V_0/\log_e(2h/a) \quad \text{[V]}. \tag{34}$$

Hence, for a typical value $V_0 = 12$ kV, $Q \approx 1.58$ kV. Substituting eq. (34) in eq. (30), we find that the electric field directly under the wire at the ground plane is $E_y \approx 2Q/h = 316$ V/m for $h = 10$ m, and that it increases slowly with height and decreases slowly with horizontal position from that point (see table 2.10). At head level,[19] the field is only about 330 V/m; but the actual field is increased substantially by the conductivity of the human body when a person is present. (This field-distortion effect will be discussed below). The unperturbed field depends logarithmically on the wire radius a, and large changes in wire size have only a small effect: halving the radius decreases the field by only about 8 percent. The field increases as $1/r$ with decreasing r in the near vicinity of the wire.

As applied to the electrified train with an overhead wire 20 ft. (≈ 6.1 m) above the rails charged to 11,000 V_{rms}, $Q_{rms} \approx 1.55$ kV. The electric field is $E_y \approx 507$ V/m at track level and increases to about 670 V/m at 3 m above the track — roughly head level for a person standing in the train. Inside the train, however, the grounded exterior conducting walls should provide excellent electrical shielding. (Of course, the risks involved in standing on the train track are much greater than those produced by the

Table 2.10. Peak Electric Fields (V/m) for One Wire Parallel to the Ground
Plane at a Height of 10 m

			x (m)		
y (m)	0	1	2	3	4
9	1,662	1,173	739	519	394
8	877	783	616	480	383
7	619	586	512	431	362
6	493	478	438	388	338
5	421	412	387	354	318
4	376	370	353	329	301
3	347	342	330	311	288
2	329	325	315	299	279
1	319	316	306	292	274
0	316	313	304	290	272

Note: The term y is the height above the ground plane; x is the horizontal distance away
from the wire. Here, the wire radius is $a = 1$ cm, the peak wire potential is 12,000 V with
respect to ground, and the parameters are similar to those for a 12-kV distribution line
($Q = 1.58$ kV). RMS values may be obtained by dividing by $\sqrt{2}$.

electric field!) The values calculated in table 2.11 represent worst-case
limits for earlier wiring geometries. The use of an out-of-phase feeder line
above the trolley wire would significantly reduce the electric fields at ground
level. (See figure 2.6 and related discussion, as well as the discussion
immediately following on electric fields from two-wire distribution lines
above a ground plane.)

The electric field lines produced in the single-wire configuration are
shown in figure 2.13. One important fact can be seen by this and later plots
of the electric-field lines: the electric field is always normal to the surface
when it intersects the ground plane, or any other conductor. This phenomenon is a direct consequence of the boundary conditions discussed above.

Two Wires above the Earth (Conducting Ground Plane)

Consider two oppositely and equally charged wires of radius a, spaced
horizontally by a distance d at a height h above the earth. From symmetry,

Table 2.11. Worst-Case RMS Electric Fields (V/m) near an Electrified Railroad

y (m)	x (m)									
	0	2	4	6	8	10	20	40	60	80
6	15,606	769	370	233	163	120	40	11	5	3
5	1,547	734	386	245	171	126	41	11	5	3
4	890	632	385	253	177	130	42	11	5	3
3	669	549	375	257	182	133	42	11	5	3
2	569	496	365	258	185	136	43	12	5	3
1	522	467	358	258	186	137	43	12	5	3
0	507	458	355	258	187	138	43	12	5	3

Note: The overhead wire is assumed to have a height $h \approx 6.1$ m above the ground plane, and the wire radius is taken to be $a = 1$ cm; y is the height above the ground plane, and x is the horizontal distance away from the wire. Here the wire potential is 11,000 V_{rms} with respect to ground. ($Q_{rms} = 1.548$ kV.)

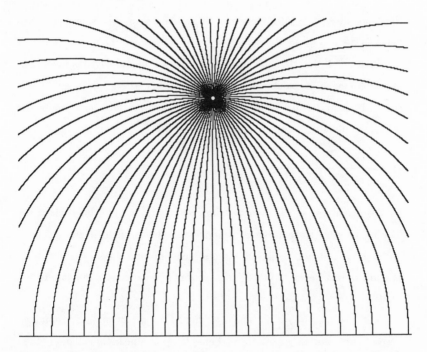

Figure 2.13. Electric-field lines produced by a single wire above a conducting ground plane

we know that an equipotential vertical plane must pass through the midpoint between the two wires and be normal to the line connecting them. If the two wires have equal and opposite potential with respect to ground, the potential on the vertical surface is zero (that is, the same as ground potential), even though that surface is not a conductor. Hence, the same type of image pair used to treat a single-wire wiring above a flat conducting plane may be used here, except that two real-line charges are in the air separated by a distance d, and two mirror-image charges are needed below ground. The problem is formally equivalent to the case of a single-line charge parallel to two conducting planes that intersect at right angles. The capacitance per unit length of one such wire with respect to ground may also be solved in closed form with complex conjugate potential theory, which yields

$$C_L = \frac{2\pi\epsilon_0}{\log_e(2hd/a\sqrt{4h^2 + d^2})} \quad [\text{F/m}]. \tag{35}$$

(Because there are effectively two capacitances of this value in series between the two wires, the actual capacitance between the two wires is half that value.)[20] If the wires are separated by $d = 2$ m and have common radii of $a = 1$ cm, the capacitance to ground for either wire given by eq. (35) is $C_L \approx 10.5$ pF/m. By writing the result in general form and combining terms, we obtain

$$Q = \frac{V_0}{\log_e(2hd/a\sqrt{4h^2 + d^2})} \quad [\text{V}] \tag{36}$$

for the magnitude of the constant that belongs in eq. (30) for each of the four image charges required to generate the electric field in the presence of the ground plane. (Adjacent charges have opposite signs.) For $a = 0.01$ m and two wires at $\pm 12{,}000$ V at a horizontal distance of 2 m and a height of 10 m, $Q \approx 2.27$ kV.[21]

As in the case where the magnetic field from a pair of wires contained currents traveling in opposite directions, there is a near cancellation of the field near ground level for the single-wire configuration. The field precisely cancels out at ground level directly under the midpoint of the two wires, and the field at head level is only about 50 V/m. The results shown in figure 2.14 and table 2.12 were computed by the method of images for two horizontal wires above a conducting ground using four line charges and should be compared to the corresponding results in figure 2.13 for the single-wire case. In contrast, the field lines for two wires in the same horizontal configuration above an *insulating* ground plane are shown in figure 2.15 (see the discussion of eq. [37] below).

Although the electric fields increase substantially near the power lines, they are relatively small near ground level. When the wire spacing d is reduced, the fields near ground level become even smaller. To illustrate this point, table 2.13 contains field values computed under the same conditions used to compute table 2.12 but for a wire spacing of $d = 0.25$ m. An example of field lines computed for two (single-phase) lines in a vertical plane above a conducting ground plane is shown in figure 2.16.

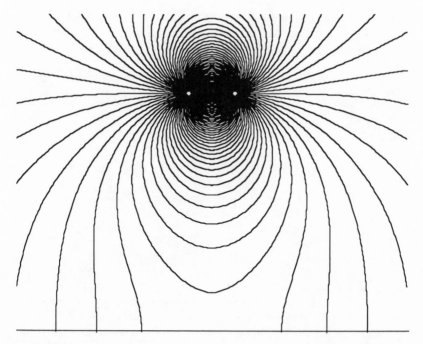

Figure 2.14. Electric-field lines from two parallel wires above a conducting ground plane (horizontal configuration)

Two Wires above an Insulating Ground Plane

If the earth were an insulator, one could better approximate the problem by computing the field from a simple pair of line charges at the wire height. Although far less realistic than the conducting ground-plane model of the earth, that case is easier to solve and gives an upper limit to the field strengths to be expected in practice. Because it has the same symmetry as the single wire above a conducting ground plane (rotated 90°), the image-charge parameter is obtained merely by replacing h with d in eq. (34). In this case, therefore,

$$Q \approx V_0/\log_e(2d/a). \tag{37}$$

The electric-field lines in this case are shown in figure 2.15 and their magnitudes are given in tables 2.14 and 2.15 for wires at quite different distances from each other but of the same height and wire radius. When we compare tables 2.12 and 2.14 for $d = 2$ m, we see that the conductivity

Table 2.12. Peak Electric Fields (V/m) for Two Parallel Wires at Equal Height above a Conducting Ground Plane (Horizontal Configuration)

	x (m)				
y (m)	0	1	2	3	4
9	2,254	2,023	1,022	503	293
8	893	793	564	367	245
7	438	497	335	256	193
6	249	239	213	181	150
5	154	151	143	131	116
4	100	100	99	97	92
3	64	66	70	73	74
2	38	42	50	58	63
1	18	25	37	48	56
0	0	17	33	45	53

Note: The term y is the height above ground; x is the horizontal distance from the midpoint between the two wires. (Wire height $h = 10$ m; wire spacing $d = 2$ m; wire radius $a = 0.01$ m; $Q = 2267$ V; and the peak wire potentials are $V_0 = \pm 12,000$ V with respect to ground.) RMS values may be obtained by dividing by $\sqrt{2}$.

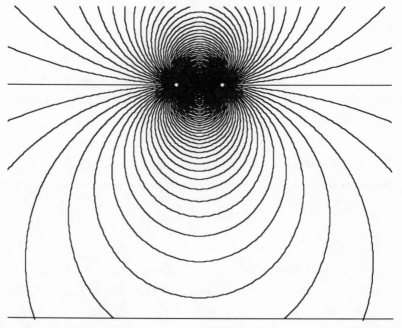

Figure 2.15. Electric-field lines from two parallel wires above an insulating ground plane (horizontal configuration)

Table 2.13. Peak Electric Fields (V/m) for Two Parallel Wires Equidistant from the Conducting Ground Plane

	x (m)				
y (m)	0	1	2	3	4
9	915	466	189	96	57
8	229	185	117	74	49
7	100	91	71	53	39
6	55	52	45	38	31
5	33	32	30	27	24
4	21	21	21	20	19
3	14	14	15	15	15
2	8	9	10	12	13
1	4	5	8	10	12
0	0	4	7	9	11

Note: The conditions used are the same as those in table 2.12, except that the wire spacing is reduced to $d = 0.25$ m. (Here, $Q = 3728$ V.) Peak potential is $V_0 = \pm 12{,}000$ V with respect to ground. RMS values may be obtained by dividing by $\sqrt{2}$.

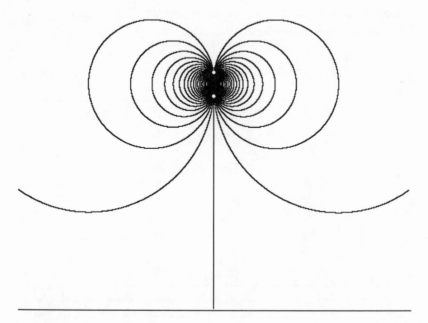

Figure 2.16. Electric-field lines from two parallel wires above a conducting plane (vertical configuration)

Table 2.14. Peak Electric Fields (V/m) for Two Parallel Wires Equidistant from a Nonconducting Earth (horizontal configuration)

	x (m)				
y (m)	0	1	2	3	4
9	2,003	1,791	896	434	248
8	801	708	497	317	206
7	401	370	299	222	162
6	236	224	194	158	125
5	154	149	135	116	97
4	108	106	98	88	76
3	80	79	74	68	61
2	62	61	58	54	50
1	49	48	47	44	41
0	40	39	38	36	34

Note: The term y is the height above ground; x is the horizontal distance from the midpoint between the two wires. (Wire height = 10 m, wire separation = 2 m, and wire radius = 0.01 m. The peak wire potentials are $V_0 = \pm 12{,}000$ V, and the line charge parameter is $Q = 2{,}003$ V.) RMS values may be obtained by dividing by $\sqrt{2}$.

Table 2.15. Peak Electric Fields for Two Parallel Wires Equidistant from a Nonconducting Earth (Horizontal Configuration)

	x (m)				
y (m)	0	1	2	3	4
9	755	383	154	77	45
8	191	153	96	59	38
7	85	77	59	43	31
6	48	45	38	31	24
5	31	29	26	23	19
4	21	21	19	17	15
3	16	15	14	13	12
2	12	12	11	11	10
1	9	9	9	9	8
0	8	8	7	7	7

Note: The conditions used are the same in all other respects as those in table 2.14. (The wire spacing is $d = 0.25$ m at a height $h = 10$ m, with a wire radius of $a = 1$ cm. The line-charge parameter here is $Q = 3{,}067$ V, and the peak potential $V_0 = \pm 12{,}000$ V with respect to ground.) RMS values may be obtained by dividing by $\sqrt{2}$.

of the earth reduces the field magnitude somewhat near ground level. But if the wires are not very far apart, the differences are not great. (Compare tables 2.13 and 2.15 for $d = 0.25$ m.)

The peak electric fields near head level seen in the above tables are much smaller than the estimates often assumed for electric fields from power lines. The numbers in tables 2.12 and 2.13 are based on exact calculations. Because the results in tables 2.11 through 2.15 are for single-phase lines, rms values can be obtained from the peak values by dividing by $\sqrt{2}$.

Three-Phase Lines

Electric-power engineers use some convenient approximate relations to treat the more complicated configurations in three-phase overhead lines. They note that if the three primary lines are placed at the corners of an equilateral triangle with a common side of length D, the capacitance to ground for each wire is given approximately by

$$C_L \approx 55.5/\log_e(D/a) \quad \text{[pF/m]}. \tag{38}[22]$$

They also find that if the wire separations are *not* constant, an effective value of the wire separation is given by

$$D_{ef} \approx (D_{12} \times D_{23} \times D_{32})^{1/3}, \tag{39}$$

where D_{12}, D_{23}, and D_{32} are the actual wire separations, and a is again the wire radius. This effective value may be used in place of D in eq. (38) to obtain a good estimate of the line capacitance to ground. On a primary wire, the rms charge per unit length is then given in terms of the rms voltage with respect to ground (V_0) by $q_L = C_L V_0$. For example, using a type 4/0 12-strand copper wire rated at 490 A with a radius of $a = 0.71$ cm in three-wire coplanar geometry with an inner-wire separation of $D_{12} = D_{23} = 1$ m yields $D_{ef} \approx 1.26$ m. Thus, equations (38) and (39) give the capacitance to ground per unit length for a representative three-phase 12-kV line as

$$C_L \approx 10.7 \text{ pF/m}. \tag{40}[23]$$

This result makes it much simpler to compute electric fields for various line configurations. Here, one can simply apply the method of images, using one charge parameter of the type

$$Q \approx C_L V_0 / 2\pi\epsilon_0. \tag{41}$$

(This equation is the same as eq. [34], except for the different phase relationships characteristic of a three-phase line.)

The common 12-kV distribution lines are largely three-phase Y systems in which the rms "phase-to-phase" voltages vary from about 12.5 to 13.8 kV. Thus the rms voltages with respect to ground (the phase-to-phase voltage divided by $\sqrt{3}$) are typically 7.2 to 8.0 kV. The *peak* voltage on any line with respect to ground is $\sqrt{2}$ times larger than the rms value, in the range between $V_0 = 10.2$ to 11.3 kV.[24] To illustrate, we shall take $V_0 = 11.0$ kV for the following examples.

The transition from peak to rms fields requires more elaborate averaging in three-phase systems than simply dividing the peak amplitude by $\sqrt{2}$. Because of the phase relationships in eq. (22), the spatial distributions of the net fields in three-phase systems vary with time. We must therefore calculate averages of E^2 over a single period in order to determine the actual rms values.

If we assume that the three-phase potentials vary as do the currents in eq. (22), the three parallel line charges will be of the form

$$Q_1 = Q \cos(\omega t), \ Q_2 = Q \cos(\omega t + 120°),$$
$$Q_3 = Q \cos(\omega t + 240°) \tag{42}$$

at median height h and spacing d, together with their mirror images $Q_4 = -Q_1$, $Q_5 = -Q_2$, and $Q_6 = -Q_3$, with the same spacing and located at distance $-h$ below the ground plane. To illustrate, the figures showing the field lines plotted here are computed at the start of the cycle ($\omega t = 0$), where

$$Q_1 = Q, \ Q_2 = Q_3 = -0.5Q \tag{43}$$

and Q is given by eq. (41).

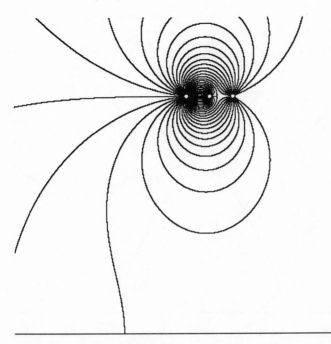

Figure 2.17. Electric-field lines for the Δ and Y systems at one instant in time (horizontal configuration). *Note:* The asymmetric field-line concentration to ground shown at the left oscillates laterally at the line frequency.

Electric-field lines for three-phase horizontal configuration lines above a grounded conducting plane, spaced horizontally by $d = 1$ m separations, are shown in figure 2.17. Field lines for the equivalent three-phase vertical configuration are shown in figure 2.18. Q_1 in eq. (39) was taken as the left wire in figure 2.17 and as the lowest wire of three in figure 2.18. The distance between the primary wires was $d = 1$ m, with the middle wire at a height $h = 10$ m. In the horizontal configuration, the asymmetric field-line concentration to ground shown at the left of figure 2.17 would oscillate laterally at the line frequency. Similarly, the asymmetric high-field intensity at the bottom of the group in figure 2.18 would oscillate up and down at the line frequency.

Values of the rms electric fields for the same configurations are given in table 2.16 for the Δ and Y systems. Here, there is no significant difference between the Δ and Y three-phase systems because there is no potential

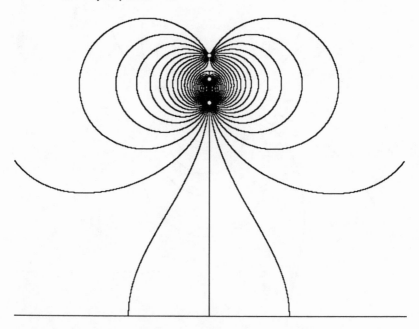

Figure 2.18. Electric-field lines for the Δ and Y systems at one instant in time (vertical configuration). *Note:* The vertically asymmetric high-field amplitude at the bottom oscillates up and down at the line frequency.

Table 2.16. RMS Electric Fields (V/m) for Horizontal and Vertical Three-Phase Systems

y (m)	x (m)								
	0	2	4	6	8	10	12	14	16
Vertical configuration									
3	69	60	41	24	13	8	7	7	7
2	59	52	36	21	10	5	6	6	7
1	54	47	33	18	8	3	4	6	7
0	52	46	32	18	7	1	4	6	6
Horizontal configuration									
3	37	40	42	39	33	27	21	17	13
2	25	31	37	36	32	26	21	17	13
1	18	25	33	34	31	26	21	17	13
0	16	23	32	33	31	25	21	17	13

Note: Line voltage (V_0) is equal to 11 kV, and the line charge parameter is $Q = 2.12$ kV, corresponding to a line capacitance of $C_L = 10.7$ pF/m see eq. [40].

difference between the fourth neutral wire in the Y system and ground. The horizontal configuration produces larger fields at ground level farther away from the pole, while the field strengths are more concentrated near the pole in the vertical configuration. The results in table 2.16 were computed for three-phase lines with a median height of 10 m and a wire separation of 1 m. As with the corresponding three-phase magnetic fields, it is important to note that the field magnitudes are determined from the vector resultant of the separate orthogonal components.

The Effect of Line Transformers

Step-down transformers used to provide 240/120-V service for homes are distributed randomly along the lines at intervals of about 100 m (one for every four or five houses). Obviously, there is substantial variation in distribution between heavily and sparsely populated areas. The most common pole transformers are connected between one leg of a Y system and ground. These single-phase transformers have power ratings that vary from 15 to 100 kilovolt-amperes (kVA) and are operated at rms primary voltages of 7.2 to 8.0 kV with respect to ground. In Y systems, the same ground (or neutral) wire is generally used for both one side of the primary line and the center tap of the balanced 240/120-V_{rms} secondary line. (With Δ systems, the primary is simply connected across the phase-to-phase voltage.)

Although these transformers have substantial capacitance to ground, this additional capacitance attached to the line does not increase the line charge. Rather, the additional charge from the transformer capacitance is localized at the transformer. This results in a field that is equivalent to the field for a point charge above a conducting plane, which falls off rapidly farther away from the transformer. The capacitance to ground at the primary terminal varies from about 3,000 pF for the most commonly used 25-kVA transformers to about 6,000 pF for those with 50-kVA power ratings.[25] (The latter are most often used in underground distribution lines, which are well shielded by the earth.)

Table 2.17. RMS Electric Fields (V/m) from a Single 25-kVA Line Transformer Mounted 10 m above the Ground

y (m)	r (m)								
	0	2	4	6	8	10	12	14	16
3	55	51	41	30	22	16	11	8	6
2	47	44	37	28	21	15	11	8	6
1	43	41	34	27	20	15	11	8	6
0	42	40	34	26	20	15	11	8	6

Note: Line voltage (V_0) is equal to 11 kV, and the charge parameter is $Q = 2.966$ kVm, corresponding to a transformer capacitance of 30 pF; y is the height above the ground, and r is the radial distance from the pole parallel to the ground.

Table 2.18. RMS Electric Fields (V/m) from Three 25-kVA Line Transformers Mounted 10 m above the Ground at 1-m Spacing Connected to the Legs of a Three-Phase Y-System Line

y (m)	x (m)				
	0	2	4	6	8
3	9	10	10	8	6
2	5	6	8	7	6
1	2	5	6	6	5
0	1	4	5	6	5

Note: Line voltage (V_0) is equal to 11 kV, and the charge parameter is $Q = 2.966$ kVm, corresponding to a transformer capacitance of 30 pF; y is the height above the ground, and x is the distance from the pole perpendicular to the wires.

The electric field may easily be evaluated by the method of images and Coulomb's law. Applying Gauss's theorem to the second of Maxwell's equations in table 1.2, the field from one point charge is

$$E_R = q/4\pi\epsilon_0 R^2 \quad [\text{V/m}] \tag{44}$$

and is directed radially outward from the charge location (the transformer) at height h above the ground, where R is the distance from the transformer to the point of observation. The image charge will be at height $-h$ (below the ground) directly under the transformer. The electric field comes from an electric dipole of moment $2qh$ and will fall off as $1/r^3$ at large distances, where r is the distance from the pole at ground level. The exact expression for the field at the point $\mathbf{P}$ is

$$\mathbf{E} = \sum_{i=1}^{2n} (Q_i \mathbf{R}_i / R_i^3), \tag{45}$$

where $\mathbf{R}_i = \mathbf{P} - \mathbf{W}_i$. The vector $\mathbf{W}_i$ points to the location of the ith charge, $\mathbf{P}$ is a vector defining a particular point in space from a common origin, and n is the number of real charges (transformers). Here, $Q_i = \pm q/4\pi\epsilon_0$ [Vm], $q = C_T V_0$, and $C_T \approx 30$ pF for the typical transformer. Thus, a representative magnitude of the charge parameter in eq. (45) is $Q \approx 2.966$ kVm for our typical peak line voltage of $V_0 = 11{,}000$ V. Values of the rms electric field as a function of distance from an origin at the base of the pole are shown in table 2.17, in which it is assumed that one transformer is located 10 m above the ground.

Equation (45) applies equally to any number of transformers, except that the electric field will no longer be axially symmetric about the pole. Table 2.18 lists values of the total rms electric field as a function of distance x perpendicular to the wires from an origin at the base of the pole, on the assumption that three transformers are located 10 m above the ground at 1-m lateral spacing on a three-phase Y system. As with the addition of the fields from wires, there is appreciable cancellation of the individual fields because of phase relationships. (Compare this with table 2.17.)

3

Natural Sources of Exposure

The Earth's Magnetic Field

At least 95 percent of the earth's magnetic field is thought to be generated deep within the earth's crust by circulating currents of uncertain origin. It is equivalent to the field produced by a magnetic-dipole moment, $M \approx 8 \times 10^{25}$ Gcm^3, which is located near the center of the earth, with an axis tilted from the earth's spin axis by about 11°.[1] Along the earth's spin axis, the field lines are primarily directed outward from the South Pole and inward toward the North Pole. That is, the field direction is close to one that would be produced by a negative charge distribution rotating with the earth about its axis. The field at any point on the earth's surface can be estimated from eq. (27), which describes the far-field components of a magnetic dipole. Over the surface of the earth, this magnetic field varies from about 0.3 G (directed horizontally) at the equator to about 0.7 G (directed vertically) at the North and South Poles. A representative value of the magnitude of the field in the United States is about 450 mG.

It seems well established from geological studies that the magnitude of the earth's field has fluctuated widely over historic, prehistoric, and geologic times. Its average value has decreased by about 10 percent since 1838 and by about 33 percent in the past 2,000 years, showing a net increase of 20 percent from 4,000 years ago. The average time between geologic magnetic-field reversals has been about 230,000 years over the past 6 million years. But changes in the value of the field of about 0.2 mG at the equator and about 0.5 mG at the Poles occur over one-day intervals, even during quiet periods of solar activity. The solar wind, consisting primarily of a flux of high-energy protons from the sun, produces (through collisions near the earth's atmosphere) about 0.2 to 30 million electron volt (MeV) protons that are trapped by the earth's magnetic field in the so-called Van Allen belts

(Hess and Greisen 1974). Because of the inhomogeneous nature of the dipole field, these particles spiral up and down the magnetic field lines, beginning at a distance of several earth radii out at the equator and plunging into the atmosphere near the Poles, where they create secondary ionization that produces the aurora borealis and alters the earth's charge distribution.[2] Hence, sudden fluctuations in the earth's field, often exceeding 100 mG, correlate with unusual solar-flare activity and the emission of large bursts of protons in the solar wind.

The geomagnetic field has a quasi-regular variation at any point on the globe of roughly 0.1 to 0.3 mG, which results both because of charged particles from the solar wind and because of the photo-ionization of molecules in the upper atmosphere by far-ultraviolet light from the sun. The vertical intensity decreases around noon in the northern hemisphere, while the horizontal component goes through maximum and minimum values during the morning and afternoon, respectively. The variations at any point on the earth are primarily controlled by the position of the sun, and there is little fluctuation at night. Two-thirds of the daily variation appears to be determined by sources outside the earth's crust and the other one-third by internal sources. The moon's gravitational force has a tidal effect on the charge density in the ionosphere, which results in a further variation in the earth's magnetic field of about 0.1 to 0.2 mG on a regular basis.

The magnitudes of the magnetic fields from power-distribution lines ($\approx$ 2 mG) — which have been proposed as a primary cause of childhood leukemia — are smaller than the earth's static magnetic field by a factor of about 250 and are roughly 1/50 the size of the sudden fluctuations in the earth's magnetic field produced by solar flares. Those fields that come from ordinary house wiring are comparable to the diurnal fluctuations in the earth's field caused by the sun and by the motion of the moon. Changes of several times the earth's present static magnetic field (including total reversal) have occurred a number of times during the period of human evolution. Altering the position of a few iron freight cars or canal barges can produce variations in the local value of the earth's magnetic field on the order of 75 mG from shielding effects alone. The time-varying magnetic-field strengths from power lines and house wiring that currently worry people are smaller in magnitude than those that you would experience by running, turning somersaults, driving a car, or even sitting still and nodding your head rapidly.[3] Although the field changes at power-line frequencies are regular and sustained over long periods of time, there has been no convincing experimental evidence that steady-state variations in magnetic fields at extremely low frequencies would have a more deleterious effect than random fluctuations of the same magnitude or that any resonant mechanisms exist in biological interactions at such low frequencies. (See further discussion in chapter 6, below.)

The Earth's Electric Field

The earth's "static" electric field is directed downward normal to the earth's surface, and reported nominal values amount to about 120 V/m (Dolezalek 1988). On the assumption that the earth's surface is a conductor, these values imply a negative surface-charge density of about 1 millicoulomb per square kilometer (mC/km^2). Variation in this surface charge, along with fluctuations in the atmosphere, presumably causes the variability of the earth's magnetic field.

This negative surface charge comes from the combination of the collisional-ionization and photo-ionization processes described above. The protons in the Van Allen belts primarily collide with nitrogen molecules, producing the N_2^+ molecular ion (whose band systems are a strong component in the aurora). The secondary electrons produced in the initial ionizing collision typically have enough energy to create a great deal of secondary ionization. These effects result in a large stream of electrons, with energies $\approx$ 1 to 10 kilo-electron volts (keV), entering the atmosphere in the Polar regions and spreading charge out through the lower portions of the atmosphere through various processes (Nicolet 1974; Akasofu 1974). The remaining positive ions, many of which undergo charge-exchange collisions with oxygen, leave a positive charge density that, when combined with the earth's negative surface charge, results in a mean electric field directed downward of about 20 V/m at an altitude of 1,400 m. Reflection from this blanket of charge in the ionosphere makes long–wavelength radio communication around the earth possible.

Diurnal fluctuations also occur in the earth's electric field, analogous to the variations found in its magnetic field, and enormous fluctuations in the ionospheric charge density correlate with sunspot activity. The latter are often significant enough to wipe out radio communication through the ionosphere. Thunderstorm activity results in local electric-field intensities of about 3 million volts per meter (MV/m) — a typical field required for ionization in the air — or more near sharp or tall protuberances at the earth's surface. The lethal properties of lightning often result from intense ground currents flowing out radially from trees, producing electrocution through enormous voltage drops in the partially conducting ground.

The ambient static electric field at the earth's surface is somewhat larger than the electric fields found under 12-kV urban power-distribution lines. Rapid fluctuations in the earth's electric field because of thunderstorm activity, however, are orders of magnitude greater than these fields. The typical lighting bolt (of which there are more than 40 million a day worldwide) requires about 3 MV/m to initiate. It transfers about 5 coulombs of charge in three to four strokes lasting 50 μs each, during which the current rises at about 10 kilo-amperes per microsecond (kA/μs), with peak currents of about 10 to 20 kA and current densities of about 1 kA/cm^2 (Orville 1974). As far as is presently known, there is no unusually high incidence of childhood leukemia among people living in areas of high thunderstorm activity. (Thunderstorms are most prevalent in warm, moist climates because of the role of convection in dipolar — or even "tripolar" — charged cloud formation.)

Methods of Field Measurement

Although elaborate and expensive laboratory instruments for magnetic-field measurement have been commercially available for some time (based, for example, on the Hall effect), such instrumentation usually requires external calibration and is seldom well suited to the measurement of fields in the mG range. A summary of commercially available equipment was recently presented by Leonard et al. (1991, table 1).

Measurement of AC Magnetic Fields

The most reliable, drift-free method to measure small AC magnetic fields is by a direct application of the Faraday effect. As may be seen by a simple application of Stokes's theorem to the first of Maxwell's equations in table 1.2, the electromotive force produced in a circular loop of wire of radius a in the presence of a magnetic field normal to the plane of the coil is

$$V(t) = 2\pi f A B_0 \cos(2\pi f t) \quad [\text{V}], \tag{46}$$

where f is the sinusoidal frequency of the field in Hz, B_0 is the magnetic-induction amplitude in tesla, A is the coil area in m^2, and $V(t)$ is the EMF in V. Winding N turns on a coil of the same radius increases the output voltage by N. A 60-Hz field of $B = 1$ mG $= 0.1$ μT, for example, applied normal to a coil of $N = 2,111$ turns wound on a radius of 2 cm would produce an open-circuited peak amplitude signal of $V = 0.1$ mV, a level that can be detected with a good oscilloscope. This method requires only voltage calibration of the measuring equipment in the case of a sinusoidal field of known frequency.

Several precautions should be taken in making measurements with such a coil. First, it should be enclosed in an electrostatic shield made from a nonferrous conductor like copper. Wiring to the coil housing should avoid ground loops, which give spurious signals picked up from magnetic fields on the coil cable. The latter can be accomplished by using a twisted-wire pair that is inside an insulated cable containing a braided electrostatic shield with only one electrical ground connection at the oscilloscope end of the lead. The twisted-wire pair then connects to the coil at the other end, and only the braided shield is connected to the copper housing around the coil. This type of wiring is standard for professional microphone-cable connections provided by audio-equipment dealers in order to avoid spurious 60-Hz pickup.

Second, the frequency of the AC field must be known. This can easily be determined for a pure sinusoidal waveform with an oscilloscope or a frequency meter. If, however, the AC field is a nonsinusoidal periodic field, the spectral composition must be sorted out with a spectrum analyzer. As discussed in connection with figure 1.3, a complex periodic waveform can introduce considerable error in field determination, owing to frequency factors that enter because the Faraday effect depends on the derivative of the flux. With a spectrum analyzer, one can determine the field at each frequency component of the waveform. Alternatively, a tunable narrow-bandpass frequency filter could be used to measure the field, one frequency at a time.

In addition, a high-impedance probe like an oscilloscope or an electronic voltmeter should be used to measure the coil voltage. A multiturn coil that is small enough is apt to have significant internal resistance, and the open-circuit voltage across the coil is what is actually determined by the Faraday effect. In practice, a high-to-low-impedance converter, consisting of an emitter follower or a one-stage operational amplifier, provides adequate isolation for low-impedance circuitry.

Finally, the plane of the coil needs to be rotated in various directions during measurement to determine which direction of the surface normal gives the maximum reading; hence, not only the maximum field amplitude but also the field direction must be known. In the case of elliptically polarized fields that arise from three-phase lines, one should measure the orthogonal components of the fields simultaneously to find the resultant magnitude of the field vector. The maximum and resultant fields differ by about 41 percent in the case of circular polarization.

Although it is straightforward to assemble the equipment necessary to make AC magnetic-field measurements, the cost may well be comparable to that of purchasing recently developed commercial instruments. Portable three-coil magnetometers incorporating microprocessors to determine resultant fields ranging from about 0.1 mG to 1,000 mG that conform to speci-

fications set by the Electric Power Research Institute (EPRI), have recently become available from several manufacturers at modest cost.[4]

Measurement of AC Electric-Field Strengths

Electric fields can be measured with a pair of rounded electrodes of known separation that are monitored by a high-impedance voltmeter. The principal practical problem with making these measurements is that the electric field becomes distorted by the presence of the human body. As a consequence, electric-power linesmen usually make their measurements by holding the meter at the end of a long insulating (plastic) rod. A shielded-probe design for this type of measurement was described by Kaune and Forsythe (1985, 247–64, especially fig. 2 and associated discussion). High-to-low-impedance converters of the type described above for magnetic-field measurement could also be used in electric-field measurements. An IEEE standard for the measurement of both electric and magnetic fields from power lines includes sources of error and calibration procedures.[5]

4

The Coupling of Electromagnetic Fields to the Body

Interaction of Fields with the Human Body and Biological Materials

Although precise calculations of magnetic and electric fields from wiring configurations in free space can easily be made using basic principles (the main uncertainty being the accuracy with which the wiring configurations are known in the first place), the coupling of these fields into the human body is another matter. The body is a complex mixture of various kinds of electrolyte and tissue, some of which are both nonisotropic and nonlinear. The dielectric constants and conductivities vary considerably with the kind of tissue and the frequency, and cell size ranges from spheres of about 1 μm in diameter, characteristic of white blood cells, to the long slender cylindrical nerve cells in the spinal column of about 1 m in length.

Under these circumstances, approaches that will provide a useful a priori analysis of the present problem are limited.

(1) We can make approximate estimates of the magnitudes of the internal fields produced by known external fields using measured values of the dielectric constants and conductivities for primary components of the body.

(2) We can estimate the unavoidable noise levels in internal fields that arise from basic thermodynamic processes.

(3) We can note the field magnitudes that have been applied in controlled biological experiments to produce observable effects.

In comparing the above quantities, we should note that for controlled experiments to be successful, the fields from (2) obviously have to be much lower than those from (3). Thus, an observation that the fields from (1) are considerably smaller than those from (2) makes a much stronger argument against the consideration of ELF fields as a significant cause of cancer than the mere assertion that the fields from (1) are smaller than those from (3). I should also point out in connection with (3) that the observation of a biological effect does not necessarily imply that the field is a carcinogen. We shall consider these three questions separately in more detail.

Basic Properties of Dielectrics

Consider a parallel plate condenser containing plates whose size is large in comparison with their separation (d). If a voltage V is applied between these plates, an electric field $E = V/d$ is produced between them that is approximately normal to the parallel conducting surfaces. (We shall neglect the fringing field near the edges of the plates.)

When an electric field is applied to a neutral conductor, the free electrons arrange themselves over the surface so that the conductor is at a constant potential and the field inside the conductor is zero. From the continuity conditions in eqs. (10) and (11), we know that the tangential component of the field is zero at the boundary, and the free-surface charge density is given in terms of the normal component E_N of the external field by

$$\rho_s = \epsilon_0 E_N = \epsilon_0 V/d, \tag{47}$$

where we have assumed there is a vacuum between the plates, and ϵ_0 is the permittivity of free space. Equation (47) has been written for the plates of the condenser, but a similar surface-charge density would arise on the surfaces of any highly conducting test bodies placed between the plates. In the discussion below, I regard the field between the condenser plates as the applied external field.

Unless the external field is already in the right direction, the fulfillment of the boundary conditions of eqs. (10) and (11) involves some distortion of the field direction when a conducting test body is introduced in the external field. The magnitude of the field is altered regardless of the direction of its surface normal.

When the external electric field is applied to a nonconducting dielectric, it induces a small displacement of the bound charges in the dielectric to produce a polarization **P** (dipole moment per unit volume), which is usually in the direction of the internal electric field and proportional to the magnitude of the field. In this case, the electric field is not zero inside the medium, and there will be both tangential and normal components of the field at the surface, in accordance with the boundary conditions in eqs. (10) and (11). Because of the displaced bound charge, the total electric field that would have been produced by the free charge on the condenser plates is reduced inside the dielectric. If, for example, the dielectric is in the form of a large conducting slab with parallel surfaces just filling the condenser, the field inside the dielectric is reduced to

$$E'_N = E_N/\epsilon_r = V/\epsilon_r d, \tag{48}$$

where $\epsilon_r = \epsilon/\epsilon_0$ is the relative permittivity, or dielectric constant, of the material. That is, an additional charge density is induced on the plates, which for low voltages is proportional to E_N and results in a total surface charge given by

$$\rho_s = \epsilon_0 E_N + P = \epsilon_r \epsilon_0 E_N. \tag{49}$$

The total charge on the condenser plate is then given by $Q = CV = \rho_s A$, where A is the plate area and the capacitance is increased to

$$C = \epsilon_r \epsilon_0 A/d. \tag{50}$$

Permittivity and Conductivity of Biological Materials

Most biological materials also have appreciable conductivity. Although considerable data are available at radio- and microwave frequencies for the conductivity and permittivity of biological tissue (Foster and Schwan 1986, 27–119, esp. 88–89),[1] relatively little has been published on the frequency range that I am considering in this book. Representative values of the relative permittivity of tissues are shown in table 4.1. As we can see, the permittivity varies only slightly with frequency over the range discussed here. Representative values of conductivity are given in table 4.2. As with

Table 4.1. Relative Permittivity for Biological Materials

	Relative Permittivity of Tissue (ϵ/ϵ_0) at Specific Frequencies			
Material	10 Hz	100 Hz	1 kHz	10 kHz
Blood			2,900	2,810
Bone		3,800	1,000	640
Fat		1.5×10^5	5×10^4	2×10^4
Liver	5×10^7	8.5×10^5	1.3×10^5	5.5×10^4
Lung	2.5×10^7	4.5×10^5	8.5×10^4	2.5×10^4
Muscle				
Skeletal parallel	10^7	1.1×10^6	2.2×10^5	8×10^4
Skeletal perpendicular	10^6	3.2×10^5	1.2×10^5	7×10^4

Source: Adapted from Foster and Schwan (1986), 89

Table 4.2. Conductivity for Biological Materials

	Conductivity of Tissue (S/m) at Specific Frequencies				
Material	0 Hz	10 Hz	100 Hz	1 kHz	10 kHz
Blood	0.67[b]		0.60[a]	0.68[a]	0.68[a]
Bone			0.0126[a]	0.0129[a]	0.0133[a]
Cell					
Fluid	0.5[b]				
Membrane	10^{-5}–10^{-7}[b]				
Fat	0.04[b]			0.02–0.07[a]	
Liver	0.14[b]	0.12[a]	0.13[a]	0.13[a]	0.15[a]
Lung	0.05[b]	0.089[a]	0.092[a]	0.096[a]	0.11[a]
Muscle					
Skeletal parallel		0.52[a]	0.52[a]	0.52[a]	0.55[a]
Skeletal perpendicular		0.076[a]	0.076[a]	0.08[a]	0.085[a]

Sources: Data from
[a]Foster and Schwan (1986), 88
[b]Barnes (1986a), 102

the permittivity of living tissue, there is only a slight variation of conductivity with frequency over this range.

The Coupling of Low-Frequency Electric Fields to Conducting Dielectrics

As noted by Polk (1986, 5–7), the relative values of the conductivity and permittivity of biological tissue and of air are such that external ELF electric fields are always normal to the surface where they enter the body, and the magnitude of the internal field is always many orders of magnitude smaller than that of the external field. The result of Polk's argument is elegant in its simplicity and comes about as follows.

Consider an interface between air (medium 1) and living tissue (medium 2). The application of the boundary conditions of eqs. (11) and (14) results in relations between the normal components of the electric field across the boundary

$$\epsilon_2 E_{2N} - \epsilon_1 E_{1N} = \rho_s \tag{51}$$

and

$$\sigma_2 E_{2N} - \sigma_1 E_{1N} = j\omega\rho_s, \tag{52}$$

where I have made use of the fact that the normal components of current density are related to the normal components of the electric field through the conductivity ($J_{iN} = \sigma_i E_{iN}$ for $i = 1, 2$) and have introduced complex notation to describe the sinusoidally dependent current and charge densities. That is, I assume that the surface-charge density varies as the real part of

$$\rho(t) = \rho_0 \exp (j\omega t) \tag{53}$$

at steady state, where $j^2 = -1$. The solution of eqs. (51) and (52) for E_{2N} and E_{1N} yields

$$E_{2N} = \frac{\sigma_1 + j\omega\epsilon_1}{\sigma_2 + j\omega\epsilon_2} E_{1N} \tag{54}$$

for the normal component of the internal field; the tangential components of the electric field are continuous across the boundary (see eq. [10]).

Next, we need approximate values for the conductivity and permittivity of the two media: for air, according to Irabarne and Cho (1980, 134),

$$\sigma_1 \approx 10^{-13} \text{ S/m} \quad \text{and} \quad \epsilon_1 \approx \epsilon_0 \approx 10^{-11} \text{ F/m.} \tag{55}$$

For living tissue, representative values in tables 4.1 and 4.2 at 60 Hz are

$$\sigma_2 \approx 0.5 \text{ S/m} \quad \text{and} \quad \epsilon_2 \approx 10^{-9} \text{ to } 10^{-5} \text{ F/m.} \tag{56}$$

Putting these values into eq. (54) for a frequency of 60 Hz, we see that

$$\sigma_2 \gg j\omega\epsilon_2 \quad \text{and} \quad j\omega\epsilon_1 \gg \sigma_1, \tag{57}$$

where $\omega = 2\pi f \approx 377 \text{ s}^{-1}$; hence

$$|E_{2N}/E_{1N}| \approx \omega\epsilon_1/\sigma_2 \approx \omega\epsilon_0/\sigma_2 \approx 0.7 \times 10^{-8}. \tag{58}$$

Thus, the magnitude of the internal field is typically smaller than that of the external field by about 10^{-8} at 60 Hz, and the coupling is roughly proportional to the frequency.

That the external field lines are normal to the surface of the tissue (that is, $\theta_1 \approx 90°$ in the following equation) may be seen by noting that

$$\tan\theta_1 = E_{1N}/E_{1T} \approx 10^{+8}E_{2N}/E_{2T} = 10^{+8}\tan\theta_2. \tag{59}$$

To take an example, at 60 Hz, even if the direction of the internal field is $0.1°$ (nearly parallel to the surface), the external field makes an angle of $89.9997°$; it is thus within $0.0003°$ of the surface normal. For 60-Hz electric fields in air, therefore, the body acts as if it were a nearly perfect conductor with a coupling coefficient of about 10^{-8}. As shown in the next section, the conductivity of the body distorts the ground plane and substantially raises the external field. This distortion can increase the effective transmission coefficient of the fields from the air by about two orders of magnitude.

The coupling of high-frequency fields is much more efficient. At 1 MHz, for example, the transmission coefficient through the air-tissue surface is ≈ 0.02 for the parallel component of the electric field (Polk 1986, 11).

Distortion of the External Electric Field Due to Coupling

The ELF electric-field coupling process is dominated by the conductivity of body tissue (σ) and the dielectric permittivity of air ($\approx \epsilon_0$). Consequently, the field lines are distorted in the vicinity of the body so that they are approximately normal to the skin. Hence, at these frequencies the presence of the body below a power line is equivalent to the addition of a protrusion of conducting material above the ground plane, with the result that the actual electric field in the vicinity of the body is both perpendicular to the skin and substantially increased with respect to its unperturbed value.

Because the unperturbed field in the air below a distribution line near ground level is nearly perpendicular to the ground and is approximately constant over the dimensions of the body, one can estimate the size of the effect by calculating the change in a uniform electric field that would be produced if a suitable bump of conducting material were placed on the ground plane. In practice, this process can be carried out analytically only when simple shapes are used for the protruding conductor.

The problem of a horizontal infinite half-cylindrical conducting boss on a horizontal conducting plane with the applied field E_0 normal to the plane (in which the field at the top of the cylinder is increased to $2E_0$), or a similar problem involving a hemispherical conducting boss (in which the field at the top of the sphere is increased to $3E_0$), is solved in most textbooks on electromagnetic theory.[2] But these solutions provide unrealistic models for the present purposes, and they underestimate the distortion by more than an order of magnitude.

In order to provide realistic quantitative values for this distortion, calculations that are more refined have been performed. Thus, Barnes et al. (1967) have shown that a prolate spheroidal conducting boss, with its semimajor principal axis a parallel to the field, results in an increase to $\approx 50E_0$ at head level for $a/b = 10$, where b is the spheroidal radius. Using the method of moments, as applied to a model in which various body parts are approximated by conducting cylinders, Spiegel (1977) presents numerical calculations based on a person standing underneath a 765-kV trans-

mission line located 14.9 m above the ground and concludes in this particular case that the fields at head level are in excess of $\approx$ 110 kV/m. But
these results are not easily scaled to the lower voltages and dimensions of
the more common 12-kV distribution lines.

Although the prolate spheroid provides a reasonable first approximation,
it is of interest to analyze the problem using proportions more typical of
human beings. I shall therefore consider a model based on a conducting
ellipsoid with the major axis aimed in the direction of the field. This problem
is about the most complex one that can be solved exactly in closed form.
Since the details of the calculation are tedious (Bennett 1992), I shall
merely outline the method and give the final solution here, which is both
very simple and quite general.

The first step is to solve Laplace's equation for the potential in ellipsoidal
coordinates and then add a particular solution that corresponds to the
uniform electric field. Next, evaluate the electric field by computing the
gradient of the potential in ellipsoidal coordinates (designated ξ, η, and ζ
by convention).[3] Finally, determine the internal field for a uniform ellipsoid
of finite conductivity and estimate departures from the ellipsoidal model
that are due to such constrictions as are found at the neck and ankles in a
real human being.

Let the height of the person equal the semimajor axis a of the ellipsoid.
(In what follows, $a > b > c$, where b and c are the other two semiprincipal
axes.) We take the field to be in the x direction along the major principal
axis and assume that the ellipsoid is sliced in half at the midpoint of the
major axis, where it rests vertically on the conducting plane. The width of
the individual (shoulder-to-shoulder distance) is then the full intermediate
principal axis ($2b$), and the thickness (chest-to-back distance at the sternum)
is the full minor principal axis ($2c$). The external field lines computed from
the general solution are shown in front and profile views of such an ellipsoid
in figure 4.1.

The maximum field is at the top of the ellipse (height of head = a) and
is given by

Front View   **Profile**

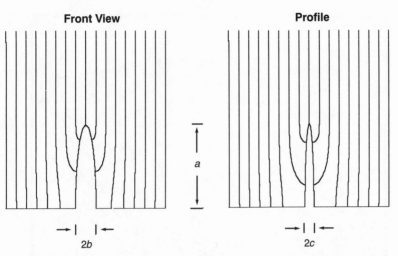

2b   2c

Figure 4.1. Electric-field lines for a conducting ellipsoid resting on a conducting ground plane in the presence of a vertical field. *Note:* The values of a:b:c used are for the two-year-old boy in figure 4.2.

$$\frac{E_{max}}{E_0} = \frac{2}{I_0 abc},\tag{60}$$

where I_0 is the integral

$$I_0 = \int_0^\infty \frac{d\xi}{(\xi + a^2)R_\xi}\tag{61}$$

with $R_\xi = \sqrt{(\xi + a^2)(\xi + b^2)(\xi + c^2)}$. Equation (60) depends only on the ratios of a:b:c. For a comparison of various subjects, let us normalize the height to $a = 1$ in the following calculations. In the limiting case of a unit sphere, where $a = b = c = 1$, the integral has the simple rational value $I_0 = 2/3$. In more general cases, the integral may be expressed in terms of tabulated, complete elliptic integrals of the second kind.[4] Nevertheless, it is actually much easier and faster to evaluate the integral numerically using a desktop computer than to look up values in tables.[5]

The component of the field normal to the ellipsoidal surface is obtained from the gradient of the potential in the ξ direction (a direction analogous to that of the radius in spherical coordinates) and is given on the surface by

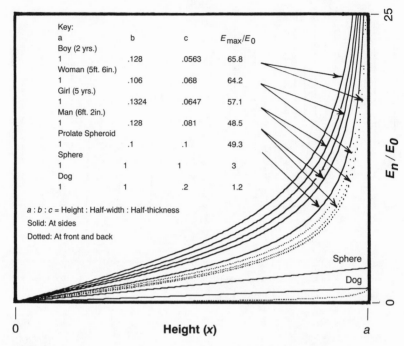

Figure 4.2. Ratio of the normal field to the applied field for a conducting ellipsoid for representative values of $a{:}b{:}c$. *Note: E_{max}/E_0* at head level is tabulated in each case.

$$\frac{E_n}{E_0} = \frac{2x}{I_0 a^2 \sqrt{\eta\zeta}} \, , \tag{62}$$

where x varies from 0 at ground level to a at the top of the head (η and ζ are the two remaining ellipsoidal coordinates). The field simplifies greatly in two particular cases:

(1) *Maximum* values of the normal field are obtained from eq. (62) at the sides of the body where $z = 0$, $\zeta = -c^2$, and $\eta = x^2(a^2 - b^2)/a^2 - a^2$ (represented by the solid lines in figure 4.2).

(2) *Minimum* values of the fields are obtained at the front (chest) and back of the body where $y = 0$, $\eta = -b^2$, and $\zeta = x^2(a^2 - c^2)/a^2 - a^2$ (represented by the dotted lines in figure 4.2). The prolate spheroid model of Barnes et al. (1967) falls close to these minimum values.

The curves in the figure are based on principal axis ratios determined for a few subjects who were conveniently at hand and are intended merely as representative cases of interest. The shape, not the height, is the predominant factor that determines the field increase. Thus, the field increase for the thin two-year-old boy is 36 percent larger than the one for the heavyset six-foot-two-inch man. At the opposite extreme, the field increases even less for the dog than it does for an infinite cylinder. The field increase at the top of the boy's head would be about fifty-five times greater than the increase at the top of the dog's head, even though the dog is only 50 percent shorter in height.

If we replace the perfectly conducting ellipsoid by a dielectric ellipsoid with a uniform conductivity ($\sigma \approx 0.5$ siemens per meter [S/m]) characteristic of the human body, some remarkable properties of ellipsoidal geometry show up. Although the external fields at power-line frequencies are unchanged from those shown in figures 4.1 and 4.2 for the perfectly conducting ellipsoid, the internal field lines now run vertically from the surface boundary to the ground plane inside the ellipsoid. The magnitude of the total current flowing in through the surface of the ellipsoid from the head (at height a) down to height x is given by

$$i_{\text{total}}(x) \approx \omega\epsilon_0 \int_x^a E_n dS \qquad (63)$$

from eq. (58) where the integral is over the surface and, surprisingly, is independent of the conductivity.

Because the dielectric is conductive, the internal field is proportional to the current density and depends on the conductivity. The average field inside the body at height x producing the current $i_{\text{total}}(x)$ flowing to ground through the cross-sectional area of the ellipsoid $\pi bc (1 - x^2/a^2)$ is downward and is given by

$$E_{\text{int}}(x) = \frac{j_{\text{total}}}{\sigma} \approx \frac{\dfrac{\omega\epsilon_0}{\sigma} \displaystyle\int_x^a E_n dS}{\pi bc(1 - x^2/a^2)_{x<a}} = \frac{\omega\epsilon_0}{\sigma} E_{\text{max}}, \qquad (64)$$

Table 4.3. Values of E_{eff}/E_0 at the Major Body Constrictions by Height

	Boy (2 yrs.) (38 in.) (96 cm)	Woman (66 in.) (168 cm)	Girl (5 yrs.) (42 in.) (107 cm)	Man (74 in.) (188 cm)	Dog (25 in.) (63 cm)
Ellipsoid parameters					
a	1	1	1	1	1
b	0.128	0.106	0.132	0.128	1
c	0.0513	0.068	0.065	0.081	0.2
Effective External Field/Applied Field					
Top of head	66	64	57	48	1.2
Neck	81	131	125	81	—
Ankles	357	488	389	357	88

Note: It is assumed that the feet make good electrical contact with the ground. Standing on one foot would double the field at the ankles for people (and quadruple it for dogs).

where E_{max} is given by eq. (60). Remarkably, for perfect ellipsoidal geometry this result is *independent* of height above the ground. For the ellipsoid, the equivalent external field is the same at all heights as it is at the top of the head.

The real human body has major constrictions at the neck and ankles compared with the ellipsoidal cross section. The current density should increase in these regions over the one assumed in the ellipsoidal model by the ratio of the ellipsoidal cross section to the real cross-sectional area, A_{real}. In calculating this ratio, we multiply the ellipsoidal area by h^2 — where h is the actual height — because the computed expressions have been normalized to $a = 1$. The effective external field is therefore increased at these locations to

$$E_{eff}(x) \approx \frac{h^2 \pi bc(1 - x^2/a^2)}{A_{Real}} E_{max}, \tag{65}$$

where $E_{int}(x) = (\omega\epsilon_0/\sigma)E_{eff}(x)$. Equation (65) was used to compile the results in table 4.3 for the subjects considered in figure 4.2.

Although the results calculated from the ellipsoidal model are exact and may easily be determined for different ellipsoidal parameters, there is no simple way to determine the differences that would arise from the more complex shape of a human being. The results for the field increase (a multiplication factor of forty-eight to sixty-six) obtained at the top of the head in the figure are significantly larger than the factor (eighteen) reported by Kaune and Phillips (1980) in a measurement done with a saline-filled dummy. The use of such manikins also involves approximations, but they are probably less severe.

Comparing the results of electric-field studies of four-legged laboratory animals that are placed between parallel condenser plates with expectations for human beings standing in the same vertical applied field has been facilitated by studies like those of Kaune and Phillips, who compared saline-filled models of humans, swine, and rats. But even with those results, if rats placed on a horizontal condenser plate in a vertical electric field were to stand up on their hind legs, they would experience an electric field at ankle level that was about two orders of magnitude larger than the one applied by the experimenter. This effect may explain some of the inconsistencies reported for such experiments.

It is assumed in the above analysis that the feet make good electrical contact with a conducting ground plane. The conductivity of the earth varies, depending on its composition and wetness, over a range in resistivity from about 1 to 10,000 ohm-meters (Ωm) (Sunde 1968, 63). Hence, the increase in external field at the body because of the coupling effect for a person wearing insulating shoes on a dry day in sandy terrain would be substantially smaller than the values in table 4.3 or in the measurements of Kaune and Phillips.

In summary, the effective electric fields can increase to at least about twenty times the unperturbed fields at head level for people standing on a conducting plane underneath power lines. Constrictions in current flow to the ground through the body at the neck and ankles might raise that factor significantly (perhaps an order of magnitude).

For a thin person standing below a typical 12-kV urban distribution line, this result implies perturbed external fields at head level on the order of 1 kV/m — that might effectively go up to as much as 10 kV/m at the ankles. For the same person standing on the steel track of an electrified railway in bare feet on a wet day, the equivalent external fields could go up to about 13 kV/m at the top of the head and increase to perhaps 100 kV/m at the ankles. If we combine these results with the calculations in the previous section (see eq. [58]), the maximum resultant internal fields coupled to the tissue are about 0.7 mV/m, a value that is nonetheless about thirty times smaller than the thermal fields that would be expected in the body at the cell level. (See eq. [71] and discussion below.)

Another way of putting these results is that the maximum effective transmission coefficient for unperturbed 60-Hz electric fields through the air-skin surface can be raised through field distortion and body constriction effects by a factor of ≈ 100 from $\omega\epsilon_0/\sigma \approx 10^{-8}$ (see eq. [59]) to about 10^{-6}.

The Coupling of Low-Frequency Magnetic Fields to the Body

Because the permeability of living tissue is close to that of free space, there is no significant distortion of the direction or magnitude of an external magnetic field as it enters the body. (See the boundary conditions of eqs. [12] and [13].) But direct interaction with an applied magnetic field is likely to be important only in the presence of permanent magnetic domains in the body. There, torque (in the case of a uniform field) and displacement (in magnetic-field gradients) can be exerted on individual magnetic dipoles. This interaction depends critically on the density of magnetic domains and pertains more to DC magnetic fields. If the density is high enough, individual domains interact with each other and can line up to form a single large macroscopic magnetic-dipole moment. This process may be relevant to direction sensing based on the earth's field in bird migration. (See discussion below.)

Electric Fields from the Lorentz Force

The effective electric fields produced inside the body that are due to the force of a magnetic field on a moving charged particle (see eq. [8]) provide a useful reference in evaluating field effects from power lines. Let me illustrate with two examples.

Blood flows through the aorta at velocities of about 0.6 m/s during systole. Hence, electric fields of about 0.6 μV/m would be generated in this flow by 10-mG magnetic fields from 60-Hz distribution lines. In contrast, the corresponding electric field in the aorta that is the result of the earth's static magnetic field would be about 27 μV/m, or about forty-five times larger. To cite an extreme case, the electric field produced by a 20,000-G magnetic resonance imaging magnet on aortic blood flow would be about 1.2 V/m.

As another comparison, the field experienced throughout the body of an astronaut traveling in an east-west orbit 200 mi. (322 km) above the earth would be about 0.4 V/m, while the field experienced by passengers flying across the country at 500 mph (806 km/h) would be about 0.011 V/m — roughly half the rms internal-field level expected from thermal noise at cell dimensions in a 100-Hz bandwidth at body temperature. (See the discussion below in connection with equation [71].) Although astronauts do experience unusual biological effects, it is generally assumed that these arise from weightlessness rather than from electromagnetic fields.

Although these examples represent DC fields that are not usually sustained for large periods of time, the electric fields generated by the Lorentz force from stray magnetic fields from AC power lines are orders of magnitude smaller.

Electric Fields from the Faraday Effect

Faraday's law of induction states that an electromotive force is induced in a closed loop of a conductor that is equal in magnitude to the time rate of change of the magnetic flux through the loop and in such a direction as to generate electric currents that produce magnetic fields that oppose the change in magnetic flux.

A useful form of Faraday's law may be derived by applying Stokes's theorem to the first of Maxwell's equations in table 1.2. Consider a circular loop of radius r, oriented in a uniform magnetic field so that the normal to the plane of the loop is in the field direction. Taking the magnetic flux to be $\pi r^2 B$, where $B = B_0 \sin 2\pi ft$, we see that the electric field around the loop is given by

$$E_{int} = -0.5 r dB/dt = -\pi r f B_0 \cos 2\pi ft \quad \text{[V/m]}, \tag{66}$$

where r is the loop radius in m, f is the cyclical frequency of the field in Hz, t is the time in s, and B_0 is the peak magnetic induction field in T (1 mG $= 0.1$ μT).

Because magnetic fields enter the body without modification, the Faraday effect could be an important way in which electromagnetic fields couple to the body. To take an example, a uniform magnetic-induction field of $B = 20$ mG rms at $f = 60$ Hz would produce an rms electric field of $E_{int} \approx 38$ μV/m over a circular loop of material $r = 10$ cm in radius. The current flowing in the loop would depend on the conductivity of the material. For a conductivity of $\sigma \approx 0.5$ S/m, representative of body fluid, the rms current density would be $J = \sigma E_{int} \approx 19$ μA/m^2. The result critically depends on the conductivity, the frequency, and the size of the loop but can be comparable in importance to the direct coupling of external electric fields from distribution lines. Moreover, the insulating properties of the cell membrane do not shield the inner portions of the cell from electric fields induced by the Faraday effect. A 20-mG, 60-Hz field would therefore result in an induced electric field of about 1.9 nV/m directly inside the membrane of a cell with a diameter of 10μm.

The Coupling of Internal Electric Fields to the Cell Membrane

So far, we have considered the coupling of external fields to produce internal electric fields in the conducting electrolyte of the body. Schwan (1988) noted that E_{int} is amplified when coupled to the cell membrane. For a specific example, consider a spherical cell of radius $r \approx 10$ μm and a membrane thickness of $d \approx 50$ Å $= 5 \times 10^{-9}$ m. From the values of conductivity given in table 4.2, it is evident that the membrane can be regarded as an insulator with respect to the tissue fluid. Solutions to Laplace's equation in this limit show that the field in the membrane will be about

$$E_{mem} \approx 1.5E_{int}r/d \approx 3{,}000E_{int}, \tag{67}[6]$$

where the angular variation of the field over the membrane is ignored. For direct coupling, all the voltage drop in going across the cell occurs across the membrane, and the membrane shields the inner portions of the cell from the applied field. (Similar shielding of the nucleus and cytoplasm by the cell membrane does not occur in indirect coupling through the Faraday effect.)

The worst-case limits estimated before for electric fields from 60-Hz, 12-kV distribution lines gave $E_{int} \approx 0.002$ V/m and correspond to $E_{mem} \approx$ 6 V/m. In contrast, the fields induced in tissue by the Faraday effect from 60-Hz power lines (≈ 40 μV/m) would result in much lower values, on the order of $E_{mem} \approx 0.12$ V/m. As I show in the next chapter, the fields across the membrane from thermal noise at body temperature are on the order of 300 V/m, at least fifty times larger than the worst-case limit.

The fields normally found across the highly insulating cell membranes are enormous, however. The voltage drop across the Purkinje cells in heart muscle fibers amounts to about 0.090 V (Netter et al. 1978, 15, 48), and nerve-cell membranes typically have potential drops of 0.05 V across them (Plonsey 1969). At a membrane thickness of ≈ 50 Å, the normal fields are $E_{mem} \approx 10^7$ V/m; that is, more than six orders of magnitude larger than the fields estimated in the worst-case limits that are the result of coupling from power lines.

5

Natural Sources of Noise

As noted by Barnes (1986b, esp. 130 ff.), there are unavoidable natural sources of electrical noise. To the extent that these natural sources produce electric fields that are much larger than those induced by external low-frequency AC sources, it is unreasonable to expect that such external sources (60-Hz distribution lines and the like) could be a significant cause of cancer. Barnes suggests three potentially important sources of electrical noise: Johnson noise (or resistor noise), shot noise (fluctuations in current that are due to the discreteness of electrical charge), and "$1/f$ noise." I shall consider these sources in reverse order.

It is thought that $1/f$ noise (where f is the frequency) is due to a variety of processes that evolve with time. As a general consideration, anything that ultimately ceases to function — or "burns out" after an average "life-time" — has an extremely large (or singular) noise component at low frequencies of about 1/lifetime. This certainly includes all components in the human body. Barnes notes specifically that $1/f$ noise is generated when ions flow through an orifice — which is fundamental to ion conduction through channels in the cell membrane. Although a great deal of evidence has been accumulated to show that the noise-power spectrum varies as C/f in various cases, it is far from clear how to evaluate the constant C in ion conduction channels. In practice the noise power actually varies approximately as C/f^x in semiconductors, where the exponent x ranges from about 0.8 to 1.5.[1]

Shot noise arises from Poisson counting statistics, as applied to an average current flow involving discrete charges (Rice 1944, 1945). The noise-per-unit bandwidth is frequency-independent, and the mean-square current flow is given by

$$\overline{i^2} \text{ noise} = 2qI_{dc}\Delta f, \tag{68}$$[2]

where q is the value of the charge, I_{dc} is the average current flowing, and Δf is the bandwidth. As an example of one of the larger currents flowing normally in the body, the peak internal fields across the chest associated with electrocardiogram signals[3] amount to about 0.01 V/m and imply current densities (for $\sigma = 0.5$ S/m) near DC of $J = \sigma E \approx 0.005$ A/m^2, through an area of about 0.025 m^2, or total currents of about 0.00012 A. On the assumption that this process involves the flow of singly charged ions, the rms noise current in a bandwidth of 100 Hz from eq. (68) would be i_{noise} $\approx 6.3 \times 10^{-11}$. Hence, the noise-current density through the same area would be $j_{\text{noise}} \approx 2.5 \times 10^{-9}$ A/m^2, or the effective noise fields would be $E_{\text{noise}} \approx 5 \times 10^{-9}$ V/m. This value is so many orders of magnitude below the level expected from Johnson noise at body temperature in the same bandwidth that we need not consider shot noise any further.

The most important source of noise seems to be the well-known phenomenon of thermal noise, which was discovered experimentally by J. B. Johnson (1928). This noise arises in a resistor from the irregular (Brownian) motion of electrons in accordance with the well-established concepts of kinetic theory. A quantitative theoretical treatment of thermal noise was first given by Harry Nyquist (1928). One can derive Nyquist's result simply by applying the equipartition theorem in statistical mechanics (which states that at thermal equilibrium, a system has an average energy of $kT/2$ per degree of freedom — or per quadratic term in the expression for the total energy)[4] to the total energy for a resonant inductance-capacitance-resistance (LCR) circuit of known bandwidth, with the result that the mean-square voltage across the resistor R per unit frequency interval Δf is given by

$$\langle V^2 \rangle / \Delta f = 4RkT. \tag{69}[5]$$

This result is quite general and has been checked for frequencies ranging from near DC through those in the microwave region. (The expression will break down at optical frequencies because it does not include the modification made to the equipartition theorem by Planck that heralded the start of quantum mechanics.)

Adair (1991) has applied Nyquist's formula to estimate the limiting unavoidable fields in the cell that are due to thermodynamic effects. If the resistor R is made from a cube of tissue of length d on an edge placed between the plates of a capacitor, Adair notes that

$$\langle V^2 \rangle = 4RkT\Delta f = 4(\rho/d)kT\Delta f \tag{70}$$

and the thermal electric field $E_{kT} = V_{rms}/d$, or

$$E_{kT} = (2/d)(\rho kT\Delta f/d)^{1/2}, \tag{71}$$

where the resistivity is given by $\rho = 1/\sigma$ (see table 4.2). If $\rho \approx 2$ Ωm for tissue, and $d = 20$ μm corresponding to a cubical volume the size of a cell, the characteristic noise field at cell dimensions and body temperature is

$$E_{kT} \approx 0.020 \text{ V/m} \tag{72}$$

in a bandwidth $\Delta f = 100$ Hz (say, from DC to 100 Hz).

Equation (72) is about five hundred times larger than the maximum internal fields estimated earlier for indirect coupling through the Faraday effect from 60-Hz magnetic fields generated by distribution lines. This value of E_{kT} is also about ten times larger than the worst-case coupling of external electric fields estimated from 12-kV, 60-Hz distribution lines because of distortion by the human body.[6] To put it the other way around, producing an internal field at the cell level of magnitude $E_{kT} \approx 0.020$ V/m requires an external 60-Hz electric field of about 3,000 kV/m — a field which is about 50 percent larger than the corona discharge limit. (A person subjected to an external field that large would glow in the dark.)

Two points should be made regarding this result. First, the bandwidth Δf needed in eq. (71) is not well known. On the one hand, to compare the noise produced by a field at frequency f, it is reasonable to assume that the bandwidth includes at least that frequency. On the other hand, if there is some natural biological filtering process that limits Δf to some low value like 15 Hz, then only the induced-field components within that bandwidth should be considered. (These induced fields might arise from nonlinear detection of the modulation envelope on a high-frequency carrier, however).

Second, although $E_{kT} \propto 1/d^{3/2}$, so that the noise field decreases with the square root of the volume as the volume increases, it is *the noise field at cell volumes* given by eq. (72) with which we are concerned. Thus, for example, the noise field from eq. (72) should be compared with the field induced by the Faraday effect in a loop $2\pi r$ in circumference from eq. (66) rather than a field that is $d^{3/2} = (2\pi r)^{3/2}$ times smaller. There can, however, be variation in both the size and the shape of the cell. Doubling the diameter, for instance, reduces the noise by 2.8; tripling it reduces the noise by 5.2; etc.

Thermal Noise in the Cell Membrane

As shown above, the fields induced in the cell membrane from 60-Hz, 12-kV lines in worst-case limits can approach 6 V/m. Because it has been suggested that induced ELF electric fields from power lines might cause cellular changes as a result of interaction in the cell membrane (see the review by Weaver and Astumian [1992]), it is particularly important to estimate the size of unavoidable thermal fields at the membrane level. This subject has been pursued by a number of authors, including DeFelice (1981), Weaver and Astumian (1990), and Adair (1991), whose argument I follow here.

On the assumption that the cell is spherical, the resistance is simply $R = \rho t / 4\pi r^2$, where t is the membrane thickness and r is the cell radius. If we take values of the resistivity — $\rho = 1/\sigma \approx 10^5$ to 10^7 Ωm — from table 4.2, a membrane thickness $t \approx 50$Å, and a value of the cell radius $r \approx 10$ μm, the membrane resistance ranges between $R = 0.4$ and 40 megohm (MΩ). From eq. (71), we know that the noise voltage across the membrane at body temperature in a 100-Hz bandwidth is then $V_{rms} = 0.8$ to 8 μV over the range of uncertainty in membrane resistance. We thus conclude that the noise field in a 100-Hz bandwidth is within a factor of two of

$$E_{kT} \approx 280 \text{ V/m},\tag{73}$$

where the main uncertainty is in the membrane resistivity.

The result in eq. (73) is about fifty times larger than the worst-case limits estimated earlier for direct coupling from the electric fields from 12-kV distribution lines.

Membrane Noise for Long Cylindrical Cells

Adair's model of the noise problem applies to spherical cells where it is probably reliable within factors of two or three. Doubling the cell diameter, for instance, would reduce the noise by only about a factor of three from the level in eq. (73).

Because there are nerve cells in the brain ranging in length from 1 or 2 mm to a few cm, and cells of as much as 1 m in the spinal column, the thermal-noise results should be modified for long cylindrical cells. In this case, the membrane resistance becomes $R = \rho t/2\pi r L$, where L is the length, and the other quantities are as previously defined. Resistance and rms thermal fields at body temperature for cells of several lengths that are all 20 μm in diameter are estimated in table 5.1, in which a resistivity of 10^6 Ωm, a membrane thickness of $t \approx 50$Å, and a bandwidth of 100 Hz are assumed. We can see there by comparison with equation (73) that the thermal noise in a spherical cell 20 μm in diameter is reduced by a factor of only a hundred from the noise in a cylindrical cell 1 m long with the same diameter.

Furthermore, there will be a compensatory decrease in the coupling efficiency from the internal field to the membrane when the long dimension of the cell is parallel to the internal field (as would be the case with vertical fields from power lines coupling to 1-m-long nerve cells in the spinal column). Because E_{tan} is continuous (see eq. [10]), the coupling efficiency would be reduced by a factor of about three thousand from that described in eq. (67) for the sphere. The net result in this worst-case limit is that there would be no significant gain in the signal-to-noise ratio at the mem-

Table 5.1. Resistance and Thermal Fields for
20-μm-Diameter Cells of Several Lengths

L (m)	R (Ω)	E_{kT} (V/m)
0.001	40,000	52
0.01	4,000	16
0.1	400	5
1	40	2

brane level over the ratio estimated for the sphere. The other cases (brain cells, for example) in table 5.1 would be intermediate because the continuity condition on D_{normal} (see eq. [11]) would again be involved. The spherical approximation is therefore not likely to give signal-to-noise ratios that are in error by much more than an order of magnitude.

Large Aggregates of Cells

Weaver and Astumian (1992) have suggested that the effects of thermal noise level might be vastly reduced in large aggregates of cells. On the one hand, they argue that the "amplification," by which they mean the factor in eq. (67) that relates the field in the cell membrane (E_{mem}) to the one in the surrounding electrolyte (E_{int}), can be increased by $N^{1/3}$ (where N is the number of cells in a spherical cluster) if aggregates of cells are electrically connected by gap junctions as occurs in some tissues (Cooper [1984] and Pilla et al. [1992]). If N were as large as a million, the worst-case electric field estimated above from 60-Hz distribution lines would then be raised by a factor of a hundred to $E_{\text{mem}} \approx 600$ V/m ≈ 2 E_{kT} from eqs. (67) and (73).

Weaver and Astumian also note that if the membranes of all N cells are electrically connected in parallel, their individual capacitances add. Under this assumption, they conclude that the effective thermal-noise level in the cell membrane decreases as $1/\sqrt{N}$ and that the effective signal-to-noise ratio (meaning ratio of E_{mem} to E_{kT}) increases as $N^{5/6}$. They then apply the result to aggregates of a million cells and conclude that the "signal-to-noise" ratio might increase by as much as 100,000.

Although it would be prohibitive to solve Laplace's equation in the large aggregate limit, we can estimate the amplification factor by equating the voltage drop in the field direction through the aggregate to the voltage drop through the electrolyte with the cells absent, noting that those across the cells will be primarily in the high-resistance membranes. Without gap junctions there is negligible difference between large aggregates and the single-cell result of eq. (67). But with highly conductive channels (gap junctions) through the cell membranes at each point of contact insulated from the body electrolyte, the cell membranes are approximately in parallel. Then the voltage drop across the aggregate is about equal to the voltage drop across the outermost cell membranes, and we get the Weaver and Astumian result for the amplification factor. Note that this factor depends solely on the number of cells connected by gap junctions along the field direction.

Their result for noise reduction can be obtained merely by substituting the reduced membrane resistance for the total parallel combination of cells in the Nyquist formula in eq. (70). The noise reduction does not really arise from increased capacitive filtering as implied by Weaver and Astumian. Consider each cell membrane as an equivalent parallel resistance-capacitance (RC) circuit with an internal-noise generator in the resistor having an rms voltage V_0 given by eq. (70). First note that although this equivalent circuit acts as a low-pass filter for the voltage across the membrane, the filter produces negligible attenuation in the ELF range. Typical cell membranes have a capacitance of about 1 $\mu F/cm^2$, hence, about 3 pF for a cell 10 μm in diameter.[7] Taking the membrane resistance to be 10 MΩ, the time constant for the low-frequency filter would be about 30 μs, and negligible filtering of the noise voltage would occur at power-line frequencies without an increase in the filter time constant (RC) by a factor of a hundred. (The voltage across the cell membrane would be reduced by $\sqrt{2}$ at a frequency equal to $1/2\pi RC$.) Applying Thevenin's theorem to N such circuits connected in parallel yields a single equivalent RC circuit in which the effective resistance (R') and capacitance (C') are given by

$$R' = R/N \text{ and } C' = NC, \text{ or } R'C' = RC. \tag{74}$$

That is, there would be no change in the filter time constant over the isolated cell membrane and no increased filtering action at power-line frequencies. But the noise voltages add incoherently and result in an equivalent single noise voltage given by $V_0/\sqrt{N}$.

All these results assume the gap-junction resistance (R_{jcn}) is zero, an assumption not likely to support extrapolation to a million cells. Reported measurements of gap-junction resistances between pairs of coupled cells differ widely. Loewenstein (1966) got 0.1 MΩ for *Drosophilia* salivary gland cells; Bennett et al. (1972) found 1 to 8 MΩ for reaggregated cells in *Fundulus* blastulae; Neyton and Trautman (1985) reported 8 GΩ from coupled cells in rat lacrimal glands. Using values of the gap-junction resistance from 0.1 to 10 MΩ in a computer model of long-chain aggregates with normal membrane resistances of 10 MΩ to 1 GΩ, I found that asymptotic limits $\approx \sqrt{R_{jcn} R_{mem}}$ were reached on the reduced membrane resistance of 2 to 10 MΩ. Because this is about the range in membrane resistance assumed to obtain E_{kT} in eq. (73), the reduced noise limit in such a large aggregate does not materially affect the arguments given elsewhere in this book. If the measured gap-junction resistances listed above are really orders of magnitude smaller than the single-cell-membrane resistances, then the membrane thermal noise in the single cell was simply underestimated in eq. (73).

Finally, although gap junctions in large aggregates are known to be present in such major organs as the heart and liver, they are not found in platelets and white cells in the blood stream. Hence, the cells most directly immersed in the body fluids, that have the most efficient coupling to externally applied fields, and that are most apt to be involved in a disease such as leukemia, are not found in large aggregates.

6

Observed Interactions with Electromagnetic Fields

Magnetite and Magnetic Fields

Permanent magnetic-dipole moments (generally thought to be domains of Fe_3O_4, or magnetite) have been found in living bacteria, pigeons, honeybees, dolphins, butterflies, tuna, green turtles, marine crustacea, and even humans.[1] In humans, single magnetic domains have been found in the adrenal gland in concentrations of twenty-six parts per billion by weight, or one to ten million single-domain crystals per gram (Kirschvink 1981).

It is thought in some of these cases that torques produced by the earth's static magnetic field on such magnetic domains (or on clusters of these domains) may serve as a navigational tool. The width of a single domain is only on the order of 500 Å and is thought to have a magnetic moment of $M \approx 6 \times 10^{-17}$ Am^2. In $A.$ *magnetotacticum* bacteria, however, chains of twenty-two particles have been reported to produce a total magnetic moment of $M \approx 1.3 \times 10^{-15}$ Am^2. The torque ($\mathbf{M} \times \mathbf{B}$) produced on such a chain is evidently great enough to orient the bacteria so that they swim in the direction of the earth's field and seems to account, among other things, for accumulations of north-seeking bacteria on the north side of a New England pond (Frankel 1986). But even with twenty-two domains coupled together in the bacterium, their interaction energy with the earth's magnetic field is only $\approx kT$, and the individual bacterium must become disoriented many times in its trip across the pond. In contrast, the interaction energy of one isolated domain in the human body with a field of 10 mG will be only

$$|\mathbf{M} \cdot \mathbf{B}| \approx 6 \times 10^{-23} \text{ J} \approx 0.01kT \text{ at body temperature} \tag{75}$$

(see table 1.3). It is therefore improbable that the low concentration of magnetite found in the human body could have an interaction energy with the magnetic fields from distribution lines and house wiring that would not be overwhelmed by thermal effects. The gradients from such magnetic-field sources are also much too small to result in significant translational forces. Adair (1993) analyzed the interactions between 60-Hz magnetic fields and unbound magnetite structures aligned with the earth's field (500 mG) in a model including linear velocity-dependent damping. Using Stokes's Law, he considered damping in spherical models for small systems rotating inside the cell that have a viscosity characteristic of cytoplasm, as well as for larger systems rotating in lower-viscosity fluids like those found in tissue plasma. These cases included structures with radii ranging from 0.03 to 10 μm; that is, from a single magnetosome free to rotate within a cell to a typical cell rigidly holding large magnetosome chains.' In all cases, the kinetic energies induced were negligibly small compared with the thermal energies for 60-Hz magnetic fields smaller than 50 mG. Since the induced energies vary as B^2, however, Adair's arguments do not preclude biological effects for 60-Hz fields larger than the earth's field.

Field Effects on the Pineal Gland

Because of its involvement in the control of mammalian circadian rhythms, some scientists suggest that a secretion, such as melatonin from the pineal gland, may play a pivotal role in the central nervous system's interaction with electromagnetic fields. Some studies also suggest that depressed melatonin levels may be linked to breast cancer (Tamarkin et al. 1982) and that melatonin may be a cancer retardant.[2]

Wilson et al. (1981, 1986) reported that the long-term application (about twenty to thirty days) of 60-Hz, extremely high-amplitude electric fields ($\approx$ 65 kV/m) suppressed the normal nocturnal increase of melatonin, raised the level of 5-methoxytryptophol, and lowered the level of serotonin in the pineal glands of rats. Statistically significant changes ($\approx$ 50 percent) were found between the exposed rats and those in the control group, but subsequent attempts to reproduce these results have failed. The fields chosen would approximate those experienced by a human being standing at ground level under a 765-kV transmission line. Wilson and his coworkers argue that shielding effects resulting from the presence of other rats in the exposure chambers would have decreased the effective field levels to about 39 kV/m; yet the rats in such an experiment would experience much higher fields if they stood on their hind legs. The fields are, of course, three to four orders of magnitude greater than those to be expected from 12-kV urban distribution lines and household wiring.

Reports of magnetic-field effects on the pineal gland have been somewhat controversial. Although early studies indicated that pineal activity and melatonin levels were strongly inhibited by exposure to static earth-level magnetic fields (of $\approx$ 1 G),[3] that conclusion seems by its very nature unreasonable because changes of that magnitude would occur every time the animal moved its head. Indeed, Reuss et al. (1985) found no effect at all on melatonin or N-acetyltransferase levels in the pineal glands of rats exposed to static magnetic fields as large as 0.14 T (1,400 G).

Later experiments by Lerchl et al. (1990, 1991) may have resolved the magnetic-field controversy. Comparing results obtained in the exposure of rats to fields of about 0.4 G, which were first slowly and then rapidly reversed, they concluded that the positive effects on the pineal gland occurred only under rapid changes of the field; that is, the result seemed to depend on dB/dt. They therefore concluded that the effect was due to electric fields induced in the pineal gland by the Faraday effect. Although relays were used to switch the direction of the current in their Helmholtz coils "instantaneously," the rise time of the resultant pulse under the experimental conditions was about τ = 7.25 ms and was applied at one-minute intervals. Hence, $dB/dt \approx$ 110 G/s = 0.011 T/s. Assuming that the pineal gland in the rat is about 1 mm in diameter, the magnitude of the pulsed electric field induced in the pineal gland in the rats would be $E_{int} \approx$ 2.7 μV/m. Modifying eq. (72) for a bandwidth of $\Delta f \approx 1/2\pi\tau$ = 22 Hz would yield thermal-noise fields at the cell level of $E_{kT} \approx$ 10 mV/m, or about 3,500 times larger than the induced field. For the same bandwidth, eq. (73) gives a value for the thermal field that is about 10^5 times the induced field at the membrane level. Although the stimulus might have occurred in the eye or in some larger induction loop, it seems clear that the signal would still be buried in thermal noise at the cell level for cell diameters of about 20 μm.

Similar conclusions apply to magnetic-field effects on the human pineal gland, an oval structure that is about 4 to 8 mm in size (Snell 1978), which has an area sixteen to thirty-two times larger than that of the rat. A worst-case rms magnetic field of about 40 mG from a 60-Hz distribution line would result in a value of $dB/dt \approx 0.002$ T/s — smaller by close to a factor of five than the peak dB/dt values in the experiments with rats performed by Lerchl and his coworkers. The induced electric field would be about 0.6 μV/m. Here, eq. (72) results in thermal fields at the cell level that are about three thousand times the induced field. At the membrane level, they are about 10^6 times the induced field.

Because the frequencies are much higher, induced electric fields in the pineal gland from the pulsed magnetic fields from video display terminals might seem more worrisome. If the peak magnetic fields at 15.75 kHz from these terminals are really as large as those reported by Schnorr et al. (1991) — about 50 mG at the operator's position — $dB/dt \approx 0.5$ T/s and would result in induced electric fields of about 1.5 mV/m in the pineal gland of the operator. But for fields at this frequency to be important biologically, the limiting bandwidth in the Nyquist formula must be at least $\Delta f \approx 16$ kHz. A reevaluation of eq. (72) for this bandwidth yields thermal electric fields of $E_{kT} \approx 0.25$ V/m — an amount about two hundred times larger than the induced signal at the cell level. Similarly, eq. (73) results in thermal fields at the membrane level of about 6×10^4 times the induced field. Conversely, if the limiting bandwidth is smaller than 16 kHz, the signal would be attenuated. The conclusion in either case is that there would be no important effect on the operator. (Although not clearly specified, it is probable that the 50-mG fields reported by Schnorr et al. were in the region of 60 Hz.)

The interpretation of the experiments by Lerchl et al. (1990, 1991) is doubtful. The duty cycle of the induced field in their experiment was only about 10^{-4}, corresponding to average electric fields of about 3 nV/m, and it is possible that some other process triggered by the switching transients may have affected the rats. Unfortunately, facilities were not available to vary dB/dt continuously so that the fields could be sustained. That could actually be accomplished fairly easily by using a sinewave oscillator and an ordinary audio-power amplifier to drive the Helmholtz coils.

Rectification and Nonlinear Processes

Barnes and Hu (1977) noted that for many passive cell membranes, an approximate nonlinear dependence of the transmembrane current is expected from the Nernst equation of the form

$$I = I_0[\exp(V_m/\eta V_T) - 1] \tag{76}[4]$$

where I_0 and η (0.25 to 1) are constants, V_m is the voltage across the membrane, and $V_T = kT/q \approx 0.0268$ at body temperature. Hence, one expects the exponent in eq. (76) to be $V_m/\eta V_T \approx 100V_m$, which would give rise to a current-voltage characteristic similar to that of a good crystal rectifier.

Although Barnes and Hu were concerned with rf and microwave effects, AM signals have been detected in the 15- to 60-kHz region, overlapping the spectrum of fields from video display terminals encompassed by this book. Schafer (1977, 1980, 1987) demonstrated this detection process with a hearing aid designed to bypass the inner ear. According to Schafer, audio signals used to modulate the amplitude of such an electric-field carrier are detected by nonlinearities within the cerebral cortex. The detection is associated with a complete lack of binaural localization and has been tried by approximately seventy-eight individuals with profound hearing loss, fifty-nine of whom were able to detect speech by this method.[5] The voltages applied between electrodes at the temple and the chin of the individual were about 1 kV at 100 percent modulation, corresponding to electric fields of about 7 kV/m for a typical temple-to-chin separation.

It is probable that Schafer's work represents an extension to low frequencies of a phenomenon reported during World War II by radar technicians who claimed to hear radar pulses. That effect was not investigated seriously until much later (Frey 1961). Explanations based on direct rf-interaction with neurons (Frey 1962) and thermoacoustic mechanisms (Foster and Finch 1974) were offered.

The external field in Schafer's work is not applied through air but goes instead through an electrode of approximately 5 cm in diameter that is located at the temple and is insulated from the skin by a thin Teflon spacer. If we take the first medium to be the Teflon dielectric with zero conductivity, the coupling approximation summarized by eqs. (57) and (58) is once again valid, except that the frequency is now ≈ 20 kHz. The dielectric constant of the Teflon insulator cancels out in evaluating the ratio of the normal field from the electrode (E_{2N}) to that in the tissue (E_{1N}). Hence, in the present case we get

$$|E_{2N}/E_{1N}| \approx \omega\epsilon_0/\sigma_2 \approx 2 \times 10^{-6}, \tag{77}$$

and the internal field would be $E_{\text{int}} \approx 0.014$ V/m at 20 kHz. From eq. (67) we know that the coupling to the insulating nerve-cell membrane would result in still larger fields across the membrane of $E_{\text{mem}} \approx 3{,}000\,E_{\text{int}} \approx 40$ V/m. Supposing that 20-μm spherical cells have a membrane thickness of 50 Å, the thermal noise at the membrane level in a 20-kHz band from eq. (73) would be $E_{kT} \approx 600$ V/m, and would be about fifteen times larger than the signal. The voltage decrease across the membrane, $V_m \approx 20\ n$V, would also result in an exponent in eq. (76) of only $\approx 2 \times 10^{-6}$ at peak modulation amplitude. Thus, unless there is an enormously large or unusually shaped cell working as the detector, Schafer's interpretation of the result, based on noise considerations alone, appears doubtful. The large networks of dendrons and synapses in the brain may provide the volume required. But the results in table 5.1 for a single long cylindrical cell may not provide adequate noise reduction. The resistance would have to go to very low values across the membrane to give a ten to one signal-to-noise ratio. Yet the effect itself is quite real.[6]

A thermoacoustic mechanism may be involved, as in the earlier microwave-detection work. It is also possible that the large applied electric fields generate ultrasound waves through the piezoelectric effect. Schafer's interpretation could be checked in part by repeating the experiment with an ultrasound transducer and applying the electric field indirectly to the brain using the Faraday effect.

It should be noted that the internal fields in Schafer's work are comparable to thermal fields in a 100-Hz bandwidth and are much larger than any of the internal fields calculated from worst-case limits for 12-kV distribution lines and household appliances.

Healing Bone Fractures by the Faraday Effect

A number of clinical studies have reported the use of periodic magnetic-field pulse bursts to enhance the fusion of fractured bones (Bassett 1968; Bassett et al. 1977, 1981; Becker 1978; Watson 1979; Brighton et al. 1981).[7] The apparatus used typically produces peak fields of up to 20 G, with repetitive pulse bursts at 15 Hz of the type illustrated in figure 1.3. It is assumed that the method involves electric fields induced in the bone through the Faraday effect. Investigators using the method report obtaining induced electric fields on the order of 0.1 to 1 V/m and current densities of about 1 μA/cm^2. Bassett et al. (1981) state that therapeutic effects occur at field levels of 0.10 to 0.15 V/m, with fundamental repetition frequencies varying from about 15 to 67 Hz and required exposure times of about twelve hours per day. The electric fields reported are substantially larger than any of those encountered in the various examples discussed above and are many orders of magnitude greater than those expected from 60-Hz distribution lines.

Apparently, the electric fields reported have been estimated from pulse rise and decay times, for only one article (Christel et al. 1981) discusses specific results from spectral analysis. It is therefore of interest to compute values of electric fields implied by the waveform and spectra in figure 1.3. For a specific example, we will assume that the peak time-dependent magnetic field is 20 G and that it is applied to a closed circle of bone 2 cm in diameter. The application of eq. (66) gives results for the electric field at discrete frequencies ranging from 112 μV/m at 15 Hz, to a maximum of 0.0028 V/m at 12 kHz. Finally, a numerical integration over the entire spectrum from 15–20 kHz gives a value of 17 V/m for the total induced electric field. Thermal noise at the cell level in this bandwidth would be about 0.3 V/m, and we at last have a reported effect that is not swamped by noise.

Values for the conductivity of bone in table 4.2 imply corresponding current densities ranging from 145 pA/cm^2 at 15 Hz, through 3.7 nA/cm^2 at 12 kHz, to a total of about 23 μA/cm^2 over the entire spectrum. These values are compatible with (but somewhat higher than) those reported elsewhere. The main currents are probably flowing at substantially different and much higher frequencies than are expected by people using the technique, however. It is not obvious that comparisons of the relative effectiveness of different pulse shapes and repetition frequencies have taken these spectral differences adequately into account.

Various experiments have been conducted with animals and on the cell level to try to refine our understanding of the bone-healing process (Aaron et al. 1989; Goodman et al. 1983; Liboff et al. 1984; Tabrah et al. 1978). Although "window" effects at repetition frequencies of about 15 Hz have been reported, the window shifts with the pulse shape and the pulse-burst duration. The information provided is often inadequate to evaluate the actual fields produced (especially their spectral distribution), and some of the results appear mutually contradictory. About all that can be said definitely is that a clear understanding at the cell level of this possible healing process has not been obtained. It is interesting to note that the window frequencies are similar to the ones reported for calcium ion efflux in the brain. (See below.)

Resonances and Window Effects

One argument that could be made in support of biological effects induced by very-small-amplitude 60-Hz electromagnetic fields is that some sharply resonant process is involved in the interaction that, by remarkable coincidence, occurs at the power-line frequency. (This interpretation, of course, could not be applied simultaneously in the United States at 60 Hz and in Europe at 50 Hz.) It is possible in principle to make the bandwidth of the system small enough in the Nyquist formula so that thermal noise over the band becomes negligible compared with the induced electric field. But the thermal electric fields depend on the square root of the bandwidth; the bandwidth would have to be reduced by a factor of a hundred, for example, in order to decrease the noise level by a factor of ten. Decreasing the bandwidth also implies sharpening the resonance by the same factor. Because we have already taken minimal values of the bandwidth to make the noise estimates given above, still further reduction is not very credible. Although very slow variations in the permittivity and conductivity of tissue with frequency have been reported (see tables 4.1 and 4.2), nothing approaching the sharpness required to overwhelm thermal noise levels has been mentioned, nor has any plausible resonant energy-transfer mechanism been suggested that could support the existence of such high-Q ($Q \approx 10^8$) resonances. (The quality factor Q is a measure of the sharpness of the resonance with respect to frequency.) Adair (1991, 1048) has also given convincing general arguments that such sharp resonances simply cannot exist at the cell level.

Nevertheless, some strange pulse-repetition frequency and amplitude window effects have been reported.[8] These reports go back to the observation by Bawin et al. (1973) that ELF modulation of very high frequency (VHF) fields disrupts the electroencephalogram patterns in cats and were followed by reports of similar frequency-dependent windows in calcium-ion efflux from chick brains (Bawin et al. 1975). Blackman et al. (1979) and Adey (1980) next reported the existence of field strength as well as frequency windows in such experiments. For reasons discussed earlier — in chapter 1 in the section on AM and FM spectra — the applied frequencies in these experiments all fall in the VHF range and do not actually contain ELF fields, as long as the conducting system is linear. The situation was compounded when Blackman et al. (1982) reported that a 16-Hz "carrier" produced the same kind of effect as a 16-Hz modulation on an rf carrier — in seeming contradiction to an earlier study by Bawin and Adey (1976). The newer result suggests that a nonlinearity in the cell membrane of the type shown in eq. (76) is rectifying the modulation signal on the rf carrier and that in both cases the ELF field is producing the effect.

Blackman et al. (*Effects*, 1985) found both frequency and amplitude windows within the 1- to 120-Hz range. In the presence of the local earth's magnetic field (of 380 mG), enhancement was reported on the application of sinusoidally varying electromagnetic fields at odd multiples of 15 Hz, together with electric-field windows centered at about 2 and 15 V/m rms. The rms-applied magnetic fields in this work were about 0.68 G. Subsequently, Blackman et al. (*Magnetic field*, 1985) reported that 30-Hz frequencies became effective when the local geomagnetic field was changed from 380 to either 253 or 760 mG. In a later article, Blackman et al. (1988) reported enhancement at 60, 90, and 180 Hz (but not at 300 Hz). In 1990, they reported that the enhancement is not observed unless the applied AC magnetic field is perpendicular to the local static magnetic field.

Models proposed to explain the window data have varied from the cyclotron resonance mechanism suggested by Liboff (1985) to a model based on soliton dynamics by Davydov (1982) and Scott (1982), to models based on Bose-Einstein condensation (Fröhlich 1968), electromagnetically induced pressure waves (Spiegel et al. 1982), and magnetic resonance (Blackman et al. 1990). In 1992, cyclotron resonance by a combination of static and oscillating ELF magnetic fields was actually patented by Liboff et al. as a method for enhancing the growth of plants. Unfortunately, none of these models has been susceptible to experimental verification.

Although there may be some unusual nonlinear mechanism at work in these experiments, it is easier to say what is not involved than what is actually going on. One effect that clearly is not of concern is conventional cyclotron resonance. The mechanism patented by Liboff et al. might make sense for ions moving in a *vacuum* under extremely high magnetic fields, but the numbers are not even approximately of the correct magnitude to explain the remarkable plant growth-rate enhancement ($\approx$ 50 percent) that they reported for prolonged exposure to fields of about 100 G through this mechanism — even in a vacuum.

Liboff et al. applied a DC offset magnetic field of about 100 G in the same direction as the oscillating magnetic field, and the frequency of the applied field was adjusted to match the cyclotron resonance condition

$$2\pi f_c = (q/m)B. \tag{78}^9$$

As with the betatron particle accelerator, the accelerating electric field at the cyclotron frequency presumably arises through the Faraday effect. Liboff et al. adjust the applied frequency to correspond to the cyclotron frequency (or to odd harmonics of the cyclotron frequency) for the actual value of B and the desired charge-to-mass ratio q/m for the particular ion of interest (Ca^{++} or Mg^{++}, for example). Ions in a vacuum would be accelerated in phase with the field, with a tangential velocity v related to the cyclotron orbit radius r by

$$v = (q/m)rB. \tag{79}$$

If we take calcium (Ca^{++}) as a specific example (atomic weight $= 40$, $q/m = 4.8 \times 10^6$ C/kg), with $r \approx 1$ μm as a representative ion-channel radius and a magnetic field $B \approx 100$ G parallel to the ion channel, the ions could acquire maximum velocities of only 0.048 m/s before hitting the channel wall. That corresponds to a translational kinetic energy of 7.7 $\times$ 10^{-29} J or a temperature of about 3.7×10^{-6} K (see table 1.3). Things that cold have rarely ever been produced in the laboratory! Stating it another way, Ca^{++} ions at body temperature would have mean thermal velocities of about 440 m/s and cyclotron orbit radii of about 9.2 mm in this field. Hence, as a first observation, the relation between orbit radius and ion velocity differs by orders of magnitude from anything that could be expected to provide enhancement of ion-channel migration effects because of cyclotron action, even if the ions were in a vacuum.

The second major objection to this interpretation is that the ions are *not* in a vacuum, or even in a dilute gas. Rather, they move in a viscous medium that prevents them from acquiring a macroscopic acceleration. The cyclotron-orbit equation is based on the notion that the centripetal acceleration of the ion is provided by the Lorentz force in the DC magnetic field. Similarly,

Table 6.1. Ion Mobilities

Ion	μm (in m^2/Vs)
Proteins	$10^{-10} - 10^{-8}$
Na$^+$	5.2×10^{-8}
K$^+$	7.6×10^{-8}
Ca^{++}	6.2×10^{-8}
Mg^{++}	5.4×10^{-8}
Cl$^-$	7.9×10^{-8}

Source: Adapted from Barnes (1986a), 102

the heating of the ion by the Faraday effect assumes that the ion can be accelerated in its orbital velocity. In reality, ions in living tissue can be accelerated only until they make a collision, something that occurs in time intervals of less than ≈ 0.1 ps. Consequently, within a few tenths of a picosecond, they acquire a constant average drift velocity proportional to, and in the direction of, the applied electric field. The coefficient of proportionality is defined as the ion mobility μ, and the drift velocity in the presence of an electric field is given by

$$v = \mu E \quad \text{[m/s]}, \tag{80}$$

where E is in V/m. Values of mobilities for ions of the type we are discussing are given in table 6.1. The process, of course, produces heating in the tissue.

One can derive a set of coupled difference equations to describe the average motion of ions under these conditions. The tangential drift velocity must be computed from a differential form of Faraday's law and provides coupling with the radial drift velocity through the Lorentz force. The coupled equations have a complex time dependence, however, and require numerical analysis. They also are nothing like the usual cyclotron orbit equations. They could give rise to heating effects from the average ion motion and might conceivably produce strange frequency-dependent effects under the right conditions. But it is to be expected that diffusion will play an important role in the actual ion motion. The induced electric fields also appear to be much smaller than expected thermal-noise fields under the conditions described.

It can be shown that the mean-square diffusion length L in time δt from Brownian motion is related to the ion mobility μ in the same viscous medium by

$$L^2 = (\mu 2kT/q)\, \delta t \quad [\text{m}^2]. \tag{81}$$

For Ca^{++} ions in a field of 100 G, the cyclotron frequency applied in the experiments by Liboff et al. (1992) would be about 7.67 kHz. Hence, it would take the Ca^{++} ion about four cycles of the applied field to diffuse by an amount greater than the width of a 1-μm ion channel at body temperature. In contrast, diffusion through about 10 μm, a distance comparable to the full cell diameter, would occur in only one period of the 15-Hz field applied in the experiments by Blackman et al. (*Effects*, 1985). In that case, diffusion would overwhelm the ion-field interaction.

In both kinds of experiment, the magnitudes of thermal electric fields at body temperature appear to be enormous compared with the size of those that would be induced through the Faraday effect, even in a vacuum: In the work by Liboff et al. (1992), the Faraday effect would give $E_{int} \approx 1.2$ mV/m for a magnetic field of 100 G applied within a cell diameter of ≈ 10 μm at 7.67 kHz, compared to thermal electric fields of about 0.5 V/m evaluated over the same region and bandwidth from eq. (71). At the membrane level, the induced fields from eq. (67) would be about 3.6 V/m and the thermal noise fields in the 7.67-kHz bandwidth from eq. (73) would be about 160 V/m. Hence, both at the cell level and at the membrane level the thermal fields would overwhelm the induced fields.

In the work by Blackman et al. (*Effects*, 1985), the Faraday effect would result in $E_{int} \approx 0.16$ μV/m for magnetic fields of about 0.68 G applied over the cell diameter at a frequency of 15 Hz. This internal field is nevertheless much larger than the field to be expected from the direct coupling process for an external applied field of 15 V/m (see eq. [58]). By contrast, in this work the thermal fields at the cell level, over the same 15-Hz bandwidth predicted by eq. (72), are $E_{kT} \approx 0.008$ V/m. At the membrane level (which is where the calcium-efflux mechanism is supposed to occur), $E_{mem} \approx 480$ μV/m and $E_{kT} \approx 1.4$ V/m. Hence, thermal noise would be about three thousand times larger than the induced fields.

Although a statistically significant effect has been reported in these experiments, no adequate explanation of the interaction mechanism has been provided. If the field effects actually occur at the cell level, the signal-to-noise ratios are unbelievably poor. It is impossible to amplify a voltage without amplifying the thermal noise present at the input to the amplifier in the same bandwidth. It is conceivable that some nonlinear model of the interacting system could be constructed to overcome the thermal-noise limit in the same manner that the well-known "lock-in detector" permits the observation of effects that would normally be buried in the noise. But that requires a system in which the effect is modulated by a reference signal of precisely defined frequency and in which the detected signal is multiplied in a nonlinear mixing element by the same reference signal (or harmonic thereof) and is then put through a low-pass filter to produce an adequate reduction of the bandwidth. Because cell-membrane capacitances are limited to about 1μF/cm^2 $= 0.01$ F/m^2 and membrane resistivities are about 10^7 Ωm, the RC time constant across the cell membrane can be only about 30 μs, and it is not clear where the necessary low-frequency filtering would arise.[10] In the results reported by Liboff et al. (1992), the low-pass filter would need a maximum bandwidth limit of about 0.1 Hz ($Q \approx 77,000$) — equivalent to an RC-filter time constant of about 1.6 s — simply for the effect to exceed the thermal-noise limit at body temperature. In the experiments by Blackman et al. (*Effects*, 1985), the maximum bandwidth would have to be reduced to about 4×10^{-8} Hz ($Q \approx 400$ million). Weaver and Astumian (1992) suggested that unusually long time constants could arise through the reaction rates of electrically induced membrane-associated enzyme activity, but they reported no specific results.

The Lednev Model

In an attempt to explain the observation of resonances at the cyclotron frequency reported by Liboff et al. (1987) in studies of ion efflux through cell membranes or by McLeod et al. (*Resonance frequencies*, 1987; *Resonance curves*, 1987) and by Smith et al. (1987) on cell motility, V. V. Lednev (1991) proposes a quantum mechanical model. Because a number of biologists seem to be taking this model seriously, it is worth discussing in some detail.

Lednev postulates a three-dimensional harmonic oscillator in which a charge that is bound to oxygen ligands in calcium-binding proteins has vibrational levels that are very widely spaced compared with the cyclotron resonance frequency in the applied constant magnetic field. He argues that the applied constant field B_0 splits an excited energy level ω of the oscillator by an amount proportional to the field into two sublevels, ω_1 and ω_2, whose difference in energy corresponds to the cyclotron resonance frequency

$$\Omega_c = qB_0/m_q, \tag{82}$$

where q and m_q are the charge and mass of the ion and it is assumed that $\omega \gg \Omega_c$. He next adds an alternating field $B_1 \cos \Omega t$ that is colinear to the applied static field at a frequency Ω near the cyclotron resonance so that the total field is given by

$$B = B_0 + B_1 \cos \Omega t, \tag{83}$$

and he concludes that the oscillator sublevels will be frequency-modulated in the form

$$\omega(t) = \omega(1 + \chi \cos \Omega t), \tag{84}$$

where $\chi = qB_1/2m_q\omega$ is the modulation depth, and angular frequencies are used. Lednev then concludes that the amplitude of electromagnetic radiation from the two excited states to the ground state can be written in the form

$$A = A_1 \exp \left[i(\omega_1 t + \alpha_1 \sin \Omega t)\right] + A_2 \exp \left[i(\omega_2 t + \alpha_2 \sin \Omega t)\right], \tag{85}$$

where $\alpha_1 = \chi_1\omega_1/\Omega$ and $\alpha_2 = \chi_2\omega_2/\Omega$; in general $\alpha_1 \neq \alpha_2$; and the transition probability from the pair of levels will be of the form $P = |A|^2$. By making use of conventional frequency-modulation theory (see eq. [7] and related footnote), Lednev obtains

$$P = A_1^2 + A_2^2 + 4A_1A_2 \cos \Omega_c t \sum_{n=0}^{\infty} J_n(\alpha_1 - \alpha_2) \cos \Omega t, \tag{86}$$

where J_n is a Bessel function of real argument $(\alpha_1 - \alpha_2)$ and of order, n. Assuming that the frequencies Ω_c and Ω are large compared with the decay rates of the excited levels, all the cosine terms vanish in a time average of eq. (86) except for $n\Omega = \Omega_c$. Thus, the average probability becomes

$$P_{av} = A_1^2 + A_2^2 + 4A_1A_2J_n(\alpha_1 - \alpha_2). \tag{87}$$

Lednev then notes that for modulation by an alternating field, $\alpha_1 - \alpha_2 = nB_1/B_0$, and the average probability for ion transitions to the ground state of the oscillator becomes

$$P_{av} = A_1^2 + A_2^2 + 4A_1A_2J_n(nB_1/B_0) \tag{88}$$

when $n\Omega = qB_0/m_q$ but reduces to

$$P_{av} = A_1^2 + A_2^2 \tag{89}$$

for all other frequencies. Equations (88) and (89) give resonant response at harmonics of the cyclotron frequency.

Although Lednev's is an interesting and mathematically straightforward model, there are several basic problems with his physical assumptions. Four of these have been discussed in a critical review by Adair (1992). Although I cover most of Adair's points and reach similar conclusions, I shall take a slightly more pedagogical approach. The basic difficulties with the model will be apparent to physicists, but many biologists will not have gone through a formal development of quantum mechanics necessary to understand the model quantitatively. In addition, there are a few subtle points that might be missed by many physicists. One refreshing aspect of the model is that it has the virtue of being susceptible to analysis. It also seems to be the first instance in which someone has had the courage to propose using quantum mechanics to solve a problem in this seemingly classical domain.

Table 6.2. Degeneracies of the First Few Energy Levels of the Three-
Dimensional Isotropic Oscillator Used in the Lednev Model

Energy	Eigen States	Magnetic Sublevels	Degeneracy
$n = 3$	$3P$ ($l = 1$)	$m = -1, 0, 1$	10
	$3F$ ($l = 3$)	$m = -3, -2, -1, 0, 1, 2, 3$	
$n = 2$	$2S$ ($l = 0$)	$m = 0$	6
	$2D$ ($l = 2$)	$m = -2, -1, 0, 1, 2$	
$n = 1$	$1P$ ($l = 1$)	$m = -1, 0, 1$	3
$n = 0$	$0S$ ($l = 0$)	$m = 0$	1

The Perturbed Oscillator States and Radiative Lifetimes

If a three-dimensional oscillator is perfectly isotropic, the energy levels are
given from the Schrödinger equation by

$$E_n = (n + 3/2)\omega h/2\pi, \quad \text{where } n = 0,1,2,3, \ldots, \tag{90}$$

ω is the classical vibration frequency in angular units (which is assumed
to be the same along each orthogonal axis), and h is Planck's constant. The
3/2 in the parentheses is the so-called zero-point energy having to do with
the Uncertainty Principle; because it is common to all the levels, it subtracts
out in calculations of the transition frequencies. The Hamiltonian (total
energy operator in the Schrödinger equation) commutes with both L^2 and L_z
(the square of the total angular-momentum operator and the z-component of
the angular momentum operator)[11] in the absence of the magnetic field. The
eigen functions of the Hamiltonian are products of Hermite polynomials in
their radial dependence and of spherical harmonics (Y_m^l) in their angular
dependence and are characterized by three quantum numbers, n, l, and m,
where $l(l + 1)h/2\pi$ are the eigen values of L^2 and $mh/2\pi$ are those of L_z
(m takes on integer values between $-l$ and $+l$). These product eigen
functions lead to energy levels that have the complex degeneracies illustrated in table 6.2, where I have used standard spectroscopic notation to
designate the levels.

When this oscillator is placed in a magnetic field (which I take to be in the z-direction), there are two additional terms arising in the Hamiltonian that are proportional to the vector potential and the square of the vector potential. The first of these terms is proportional to L_z and provides splitting of the energy levels in the magnetic field. It commutes with the original Hamiltonian and removes the degeneracy for the different values of m associated with the same value of l and the same principal quantum number n. The second term (proportional to the square of the vector potential) would destroy the three-dimensional isotropy of the oscillator.[12] But this term is implicitly neglected in the approximation made by Lednev that $\omega \gg \Omega_c$. Hence, the oscillator problem in the magnetic field is easily solved in terms of the eigen functions of the original Hamiltonian when the field is applied in the z direction. With the field on, the original energy levels are changed by an amount

$$\Delta E = (mqB_0/2m_q)h/2\pi, \tag{91}$$

where m goes from $-l$ to $+l$ in integer steps. The levels thus split into $2l + 1$ components: the P levels ($l = 1$) split into three components, the D levels ($l = 2$) split into five, the F levels ($l = 3$) split into seven, and so on.

Because the three components of the electric-dipole operator can be written as combinations of rY_m^l where $m = -1$, 0, 1, and the spherical harmonics are orthogonal when they are integrated over solid angle, electric-dipole transitions from the ground state are connected only to the higher excited P states. No other states of the oscillator can radiate to the ground state in the electric-dipole approximation.[13] Above the $n = 1$ state, however, there is a high degree of degeneracy with other levels having the same principal quantum number, and this degeneracy will tend to make the states extremely fragile against destruction in collisions. (However, they will primarily be susceptible to destruction by electric-quadrupole rather than electric-dipole perturbations.) For that reason, I shall concentrate on the $1P$ state in the Lednev model.

Knowing the wave functions permits making an exact calculation of the radiative lifetimes for the levels. The total $1\,P$ decay rate through spontaneous emission in all polarizations on the $1\,P \to 0\,S$ transition at frequency ω is given in the electric-dipole approximation by

$$A_{10} = \omega^3 (8\pi q^2/3hc^3)|\int \psi_{1P}^* \mathbf{r} \Psi_{0S} dV|^2, \tag{92}[14]$$

where the quantities are in the cgs units normally used in atomic physics. The wave function for the ground state is

$$\Psi_{0S} = C_0 \exp(-\beta r^2), \tag{93}$$

and the three $1\,P$ wave functions are of the form

$$\Psi_{1P} = C_1 r \exp(-\beta r^2)[Y_1^1,\ Y_0^1,\ \text{or}\ Y_{-1}^1], \tag{94}$$

where $C_0 = \beta^{3/4}/\sqrt{(2\pi)}$, $C_1 = 2\sqrt{2}\ \beta^{5/4}/\sqrt{3}\ \pi^{1/4}$, and $\beta = 2\pi m_q \omega/h$. The normalization constants are chosen so that $\int |\psi|^2\, dV = 1$ for each level where the integral is over all space. The differential volume (dV) is defined by $r^2 dr d\Omega$, where r is the spherical coordinate in the radial direction, and $d\Omega$ is the differential solid angle. The square of the radial matrix element for the transition to the ground state is the same for each of the three $1\,P$ states and is given by

$$|\int \Psi_{1P}^* \mathbf{r} \Psi_{0S} dV|^2 = \sqrt{\pi}/4\beta. \tag{95}[15]$$

Putting eq. (95) into eq. (92), we find the common spontaneous radiative decay rate from each of the $1\,P$ sublevels to be

$$A_{10} = \omega^2 q^2 \sqrt{\pi}/3c^3 m_q = 1.19 \times 10^{-26} f^2 \text{ Hz}, \tag{96}$$

where the formula has been evaluated (using cgs units) for Ca^{++} ions and f is the cyclical vibration frequency of the oscillator in Hz.

Problems with the Model

There are some specific differences in detail between the actual level structure of the three-dimensional oscillator in a magnetic field and that used by Lednev in his analysis of it. The first excited state splits into three levels in the magnetic field rather than two. But the two outer levels ($m = \pm 1$) have the same splitting as the level pair Lednev describes and will serve his purpose.

It is clear from eq. (96) that the levels will not radiate at a significant rate until the vibration frequency reaches the near-optical range. If we take $hf = kT$ at body temperature (which would be required for significant thermal excitation of the 1 P levels), $f = 6.46 \times 10^{12}$ Hz (see table 1.3) and the radiative lifetime ($1/A_{10}$) of the first excited levels is about two seconds. This lifetime is about a factor of four shorter than the lifetime estimated by Adair (1992) using more approximate methods, but it is roughly equivalent. Although we can detect radiation from such long-lived levels when they are located in outer space (witness the famous "nebulium" lines of the 1920s),[16] it is nearly impossible to do so in the laboratory. It becomes especially difficult when the states are in a collisional environment provided by a gas or liquid at atmospheric pressure.

There are two distinctly different collision processes with surrounding molecules and ions that will disrupt the emission from a quantized oscillator: *inelastic collisions,* which actually destroy the state, and *elastic collisions,* which merely broaden the transition line width.

In an inelastic collision a large amount of internal energy (that is, energy coming from vibrational excitation in the oscillator) is exchanged with kinetic energy of the colliding molecules. The probability for such collisions is especially high in the present case because the particle in the oscillator has a net free charge. Hence, very-long-range interactions with other ions will arise from Coulomb forces and with neutral molecules through induced polarization forces. The magnitude of the internal energy exchanged is limited by the interaction energy between the charge and the colliding molecule at the distance of closest approach. Even with neutral molecules at body temperature, that energy is at least on the order of 0.5 eV. The cross sections for atomic state de-excitation in two-body collisions under these conditions are typically more than 10^{-14} cm^2 for energy level changes of a few kT at body temperature.[17] Collision deactivation for such large cross sections in two-body collisions would take place at the rate of about 10 GHz at a pressure of one atmosphere. Spectroscopic studies in noble gases near atmospheric pressure have shown that states with lifetimes as short as 10 ns do not have time to radiate before they are destroyed unless they are separated by more than a few eV from lower-lying levels.[18] Three-body collisions relax these restrictions on energy exchange even more because there is an extra particle to help conserve energy and momentum in the collision. Three-body collision rates increase as the square of the pressure and would probably be $\approx$300 MHz for conditions in a living cell.[19] Hence, the oscillator states described by Lednev would probably last no longer than about 0.1 to 3 ns — or about 1 billionth of the radiative lifetime of the first excited vibrational state.

In an elastic collision no internal energy is exchanged with the kinetic energy of the colliding molecules. The molecules merely bounce off in different directions so as to conserve total kinetic energy and momentum. However, the phase of the wave function of the radiating oscillator is interrupted by such a collision. Although the oscillator will still radiate, the spectral width of the radiation is broadened because of the shortened time, T_c, for sinusoidal emission between sudden changes of phase. The frequency distribution of the radiation will be spread over a band $\Delta f \approx 1/T_c$, where T_c is the time between collisions, and will have a distribution similar to that shown in figure 1.2. The time T_c is roughly equal to the reciprocal of the collision rate, which for large-angle elastic scattering between an ion and a neutral molecule involves cross sections of $\approx 10^{-15}$ cm^2. For two-body collisions at one atmosphere pressure, the spectral width of the radiation would be broadened to $\Delta f \approx 1$ GHz by elastic collisions alone. Similarly, line broadening by destruction in inelastic collisions will be of about this same magnitude.[20]

So far, I have concluded that the oscillator states will be destroyed by inelastic collisions long before they radiate and that, even if they do radiate, the line width for the resonance would be enormously broadened by collisions. These observations do not mean that the 1 P state would not be populated. Indeed, the inelastic-collision mechanism (including the inverse process) is the principal one providing a Boltzmann population distribution in the excited states of the oscillator.[21] Similarly, although the radiation line width would be astronomical compared with the spacing between harmonics of the cyclotron resonance frequency in eq. (82), it would still be much narrower than the breadth of thermal radiation at body temperature. However, the very low transition probability in eq. (96) means that thermal equilibrium in the oscillator cannot arise through interaction with thermal radiation at pressures near an atmosphere. The principal mechanism that rules out any coherent resonant process in this model is phase interruption by collisions.

Adair (1992) suggests yet two more problems: first, it is highly unlikely that a spherically symmetric oscillator would result from the ligand environment envisioned by Lednev in ion channels through the cell membrane. If the spherical symmetry is destroyed, the simple splitting of the levels in the magnetic field vanishes with it. If the vibrational energy spacing is comparable to kT at body temperature, it is also likely that the random motion of ions and molecules will destroy this symmetry itself. One can still solve the asymmetric oscillator problem by means of quantum mechanics, but the degeneracies are removed in the *absence* of the magnetic field. No two levels will coincide and the separation between two levels in the magnetic field no longer obeys eq. (82).

Second, eq. (85) implies a special relative-phase relation for the two substate wave functions. One needs to prepare the states in a well-defined way to achieve this phase relation. There is no conceivable manner to achieve that within the Lednev model, and the actual phases will be randomized by collisions within about 1 ns. Hence, as applied to cells in living plants and animals, the model proposed by Lednev (1991) is unreasonable.

7

Questions for Future Research

Because the wavelengths of ELF electromagnetic fields are enormous compared with the dimensions with which we are concerned, electric and magnetic fields may be determined in a straightforward manner in the absence of the human body from Maxwell's equations for most cases involving electromagnetic fields from power lines. Only the "near" fields close to the source are of importance. These fields typically fall off at rates of between $1/r^2$ and $1/r^3$ at large distances r from the source. Radiation from these fields is negligible, and it is neither meaningful to specify the fields in terms of radiant power — as is frequently done with ionizing radiation at much higher frequencies — nor appropriate to try to relate them through the impedance of free space. Maxwell's equations separate in the ELF range, and the electric and magnetic components can be calculated independently.

The main difficulty in determining fields from power lines arises from the variability of spatial configurations. Figures 7.1 and 7.2 show representative worst-case limits of the magnetic and electric fields computed near head level for typical 12-kV urban distribution lines carrying currents of about 500 A. The fields from three-phase lines were computed on the assumption of a 1-m wire separation at a mean height of 10 m. But in most urban environments a wire-spacing of less than 1 m is used; the fields computed near ground level therefore represent conservative upper limits.

The worst-case magnetic fields for three-phase 12-kV distribution lines are based on an assumed 50 percent unbalanced current flow through the primary wires in Δ and Y systems. This represents a degree of unbalance that power companies would not tolerate under normal circumstances. Direct measurement in a variety of situations indicates that representative magnetic fields are about an order of magnitude smaller than the worst-case limits shown in figure 7.1. The shaded areas in the figure present the typical range

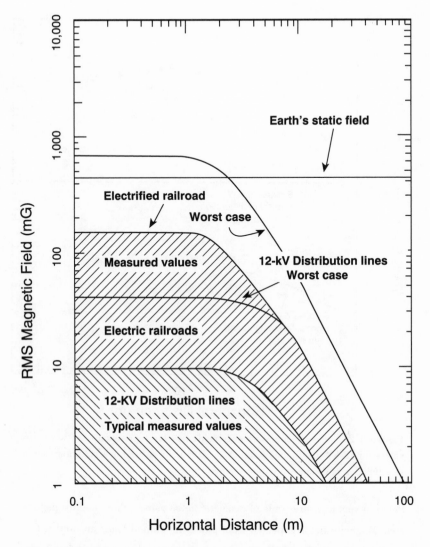

Figure 7.1. RMS magnetic fields at head level for electric railroads and three-phase, 12-kV urban-distribution lines. *Note:* The solid curves represent calculated worst-case limits; the shaded curves represent the typical range in measured values obtained with a three-coil magnetometer at 60 Hz.

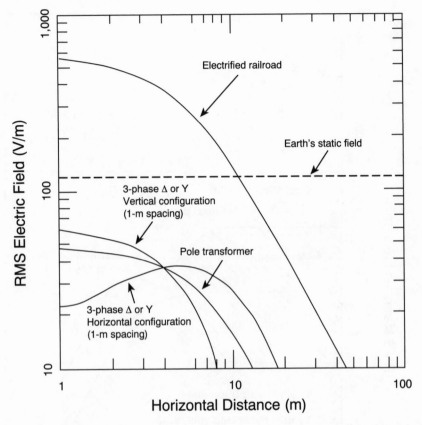

Figure 7.2. Summary of rms electric fields that are representative of worst-case values for 12-kV distribution lines. *Note:* The fields at head level (2 m) are shown as a function of distance from the source.

found in spot measurements of magnetic fields using a three-coil magnetometer. Magnetic fields expected from plumbing ground loops are well within the shaded region for 12-kV distribution lines, and those from modern house wiring (Romex or BX cable) are too small to be included in the scale of the drawing. Differences in the spatial variation of electric fields between vertical and horizontal three-phase wiring configurations are shown in figure 7.2, along with the electric fields produced at head level from a single pole transformer. (The fields from clusters of two or three pole transformers are smaller.)

As these figures illustrate, both the magnetic and electric fields from typical three-phase 12-kV distribution lines are substantially smaller than the corresponding values of the earth's field. Worst-case magnetic fields are about 40 mG, and worst-case electric fields are about 60 V/m. The fields near electrified railroads, however, can be much larger. The reason for this difference is clear. The largest fields in all instances fall between the wires carrying opposing currents or charges. All the wires for the three-phase 12-kV distribution lines are suspended in the air (or buried underground) and thus are at a substantial distance from a person standing near them. By contrast, in the case of the electrified railroad, head level lies between the overhead wire and the rails. Worst-case limits in this instance correspond to magnetic fields of 700 mG and electric fields of 600 V/m and are by far the highest fields encountered in the ordinary urban environment over extended distances. But the worst-case limits shown here for electrified railroads correspond to an early wiring geometry that does not include the use of out-of-phase feeder lines, and both the electric and magnetic fields from post–1986 wiring configurations on commuter lines can be smaller by significant factors than the worst-case limits in the figures. Although peak magnetic fields of about 300 mG at 60 Hz are presently encountered inside electrified Amtrak trains, they arise from short bursts of power and result in typical average fields of about 35 mG. Peak fields of about 650 mG, with average values of about 125 mG, were encountered on Amtrak trains in areas using older wiring configurations at 25 Hz. (See figs. 2.6 and 2.8 and related discussion.) People standing on a station platform or working along the tracks encounter both fields. Passengers in metal cars are well shielded from the electric fields, although the stainless-steel cars used by Amtrak do not produce significant shielding from magnetic fields.

Measured values of the magnetic fields for a variety of ordinary household appliances are shown in figure 7.3. Although some of these appliances produced the highest magnetic fields encountered in the present study, the values of these fields all fall off as $1/r^3$ beyond distances on the order of 0.1 to 1 m and do not represent significant sources of prolonged exposure in most instances. Because most household appliances (motors, refrigera-

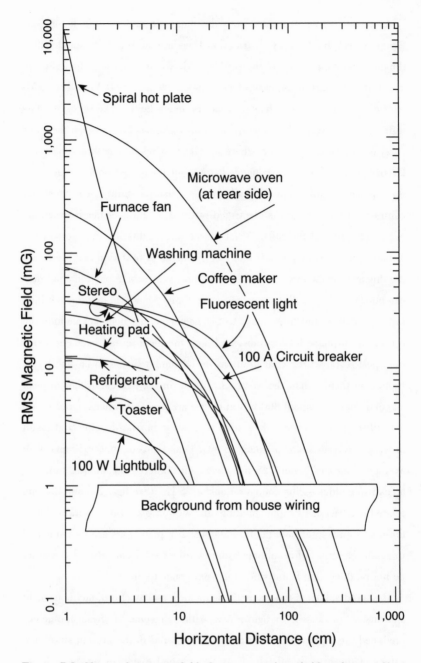

Figure 7.3. Measured magnetic fields from common household appliances. *Note:* Calculated values are shown for the spiral hot plate because the spatial resolution of the meter available was inadequate to determine such rapidly increasing fields at small distances. (Also see table 2.8.)

tors, and the like) that might produce large 60-Hz electric fields are enclosed in grounded metal housings, electric fields from such sources are not expected to be too significant with regard to health.

The problem of field estimation becomes more complicated when a person is located on the ground beneath a distribution line. Because of the high conductivity of biological tissue at ELF in comparison with air, the electric fields can be greatly increased near the person. Models in which the human body is replaced by a saline-filled manikin show that the perturbed field at the top of the head would increase by at least eighteen times. This limit assumes that the person makes good electrical contact with the ground and that the ground itself is highly conductive. Thus, for example, someone standing barefoot on a wet railroad track might expect perturbed electric fields as high as 10 kV/m at head level, whereas a person wearing insulating shoes placed on dry sand would encounter much lower fields. This effect is roughly independent of height and depends primarily on the proportions of the body.

Whether these fields are dangerous, however, depends on biological interactions. It is expected that the only important cancer-producing biological effects would occur through the interaction of humans with electric fields at the cell, or cell-membrane, level. The absence of permanent magnetic dipoles, let alone any significant mechanism of interaction by 60-Hz fields with them, eliminates the need to discuss direct magnetic interactions.

The coupling of the perturbed electric fields through the air-skin boundary introduces another level of complexity to the problem of analyzing the fields. Because of the discontinuity in conductivity and permittivity at the surface of the skin, the perturbed external fields are attenuated by a frequency-dependent factor of about 10^{-8} at 60 Hz. Thus, the worst-case 10-kV/m perturbed fields at head level are reduced to internal fields of about 0.1 mV/m in the tissue electrolyte. Allowing for constrictions in the body, this worst-case number increases by another order of magnitude to about 1 mV/m at the ankles. But this is simply the field in the body electrolyte. The resultant fields across a typical cell membrane will go up by another factor of about three thousand to worst-case limits of about 3 V/m because of the high membrane resistivity.

Because magnetic fields are almost completely unaltered by human beings (except knights in suits of armor), electric fields from the Faraday effect will couple much more efficiently into the body. Taking the worst-case magnetic field of 680 mG (from electrified railroads), the induced electric field at 60 Hz over a loop in the body with a radius of 10 cm would be about 1.3 mV/m in the tissue electrolyte, increasing to about 4 V/m across the cell membrane. Hence, both worst-case limits for electrified railroads give electric fields of 4 to 10 V/m at the cell-membrane level and would be lower by about an order of magnitude in the case of three-phase 12-kV distribution lines.

Thermal noise provides the major source of unavoidable natural fields in the body and results in randomly oriented electric fields that depend on the resistivity of the tissue and that increase as the square root of the bandwidth divided by the volume. Because we are concerned with interactions at the cell level as a potential threat to health, it is the size of these fields at body temperature within the cell volume (or within the cell-membrane) that provides the limiting test of the importance of externally induced fields. At body temperature, these thermal fields amount to about 0.020 V/m at the cell level and increase to within a factor of two of approximately 280 V/m across the cell membrane in a bandwidth of 100 Hz. External electric fields in excess of the corona-discharge limit in air would be required to produce fields of the thermal-limit size at the cell level. People subjected to such large external fields would literally glow in the dark and produce static interference on AM radio receivers.

Even in the worst-case limits found for electrified trains, thermal fields are about ten times the size of the induced fields at the cell level and about thirty times the size of the induced field at the cell-membrane level. These ratios go up by still another order of magnitude for 12-kV urban distribution lines. In all cases considered, including 60-Hz fields from various household appliances and 16-kHz fields from video display terminals and television sets, unavoidable thermal fields appear to overwhelm the induced electric fields at the cell level. Therefore, nothing in the present analysis provides any sound basis for concluding that the ELF fields from household appliances, 12-kV urban-distribution lines, or electrified railroads could have any significant deleterious effect on human health.

Several points have been developed in this book that may be worth additional consideration from an epidemiological point of view. First, the 12-kV distribution-line transformers are filled with insulating oil that may have carcinogenic properties. These transformers can overheat under heavy load conditions and the oil evaporate, possibly creating health problems correlated with power substations and high-current distribution lines.

Then, no studies dealing with commuters or workers on electrified railroads have been published. Because both the electric and magnetic fields from these overhead lines proved larger at head level by orders of magnitude than those from the 12-kV urban distribution lines in the Wertheimer-Leeper study (1979), it might be desirable to determine whether there is any unusual evidence of leukemia among these groups. The commuters are primarily exposed to the magnetic fields, while people working on the railroad are exposed to both electric and magnetic fields.

It would also be desirable to resolve the more controversial observations and interpretations discussed earlier in relation to observed interactions of the fields with the human body and with biological materials. In only one of the cases discussed (healing bone fractures by the Faraday effect) was it clear that the induced electromagnetic fields were actually much greater than thermal noise at the cell level. Ironically, in this case an explanation of the healing process has not been established.

In the other instances, thermal noise at the cell and cell-membrane levels appears to be orders of magnitude greater than the experimentally induced fields in the observational bandwidth. Nevertheless, results have been reported that appear to be marginally significant statistically. In some instances simple further experimentation could clarify the issue. One could continue the work on the pineal gland begun by Lerchl et al. (1990, 1991), for example, by varying the magnitude of dB/dt continuously and thereby increasing the duty cycle of the experiment over some four orders of magnitude. Besides increasing the size of the effect, this step would overcome a potential source of systematic error arising from switching transients in the previous experiment by Lerchl et al.

The ion-efflux experiments in brain tissue have been going on since the 1980s. Marginal, seemingly contradictory results have been reported in dozens of articles and interpreted in at least half a dozen different ways (see chapter 6, above). Some of the models used to explain the data do not make much sense on a physical basis, and it seems possible that many of the experiments may involve subtle forms of systematic error. It is likely that progress in understanding the ion-efflux phenomenon will require more basic, definable experimental methods than those which have been tried to date. If there is a real window effect in both the frequency and the amplitude, it probably involves a nonlinear interaction at the cell-membrane level. Experiments designed to study the nonlinearity per se could be the most fruitful. Because the frequency windows often seem to coincide with those reported in the bone-fracture healing work, the same nonlinearity could conceivably be involved in both cases.

Finally, the work by Schafer (1977, 1980, 1987) on detection of modulated low-frequency electric-field carriers seems to involve a nonlinear effect that has eluded adequate explanation. It appears that thermal noise would mask the phenomenon if the mechanism (based on direct nonlinear detection of AM-carrier waves at the 15- to 40-kHz level) is actually the one proposed by Schafer. Some simple experiments, ranging from the use of the Faraday effect to ultrasound transducers to explore this nonlinearity, could be performed to provide a more direct check of Schafer's hypothesis and gap-junction effects in large aggregates of cells might raise the signal-to-noise levels in this and other experiments involving brain tissue.

Notes

Introduction

1. Exaggerations and misrepresentations in Brodeur's book based on these articles have been discussed in detail by E. R. Adair (1991) and Jauchem (1992).
2. The result follows from the theorem on the addition of variances in random, independent processes. The variance for the different count is equal to the sum of the two separate variances, which for Poisson processes merely equals the separate mean values. By definition, the standard deviation is then the square root of the total variance. (See, e.g., Cramér [1946].)
3. The combination of smoking and drinking alcohol causes the relative risk for some cancers to go up by another order of magnitude.
4. Such diverse types of cancer in the cluster on Meadow Street is actually a strong argument against any one specific cause in the environment. Indeed, the Connecticut Department of Health Services concluded that there was no cluster (Linscott 1991).
5. For a critical review of the epidemiological data available before June 1992, see Trichopoulos (1992).
6. See the reviews by Brent et al. (1993) and by Brady and Reiter (1992).
7. See the review by Stein (1992).
8. See, e.g., Altman (1993). A distinguishing feature of AIDS is its continued exponential growth. Without a cure or vaccine, the current rate of growth suggests that most of the population of the world will be infected by the year 2025.

1
The Nature of Low-Frequency Electromagnetic Fields

1. See, e.g., Bleil (1957).
2. Bassett et al. (1981) suggest that high-frequency components in the pulse-burst spectrum produce deleterious heating effects that can be avoided by using a single sustained pulse in each period of the waveform rather than a pulse burst.
3. Specifically:
$$E(t) = A_0\{J_0(A_m) \sin \omega_c t$$
$$+ J_1(A_m) [\sin(\omega_c + \omega_m)t - \sin(\omega_c - \omega_m)t]$$
$$+ J_2(A_m) [\sin(\omega_c + 2\omega_m)t + \sin(\omega_c - 2\omega_m)t]$$
$$+ J_3(A_m) [\sin(\omega_c + 3\omega_m)t - \sin(\omega_c - 3\omega_m)t]$$
$$+ \ldots\}.$$
4. For a variety of methods of solution, see Schelkunoff (1943), Smythe (1950), Stratton (1941), and Jackson (1975).
5. In SI units, the magnetic induction field is expressed in tesla (T), not gauss (G): $1\ T = 10^4\ G$; $1\ mG = 10^{-7}\ T = 0.1\ \mu T$. In gaussian units, there is no quantitative difference between $\mathbf{B}$ and $\mathbf{H}$ in air because the permeability is close to unity. In SI

units, however, **B** differs from **H** by a factor approximating the permeability of free space. (See tables 1.3 and 1.4.) Although many authors used to the cgs (centimeter-gram-second) system of units use the terms *magnetic field* and *magnetic induction* interchangeably, if the field is quoted in gauss (or tesla), what is meant is actually the magnetic induction, **B.** When dealing with magnetic induction fields in air, most authors find it convenient to drop the attributive adjective *induction*, allowing the units to be self-explanatory.

6. These conditions may be derived by applying Gauss's theorem and Stokes's theorem to Maxwell's equations, as applied to small volumes and surfaces encompassing the surface boundary.

2
Sources of Low-Frequency Fields

1. Although 12 kV seems to be the most common distribution-line voltage, some areas use lines varying from 5 to 15 kV.
2. See, e.g., Tell et al. (1977).
3. See, e.g., Savitz and Calle (1987), Gauger (1985); also see summaries of these and other data given by Leonard et al. (1991).
4. James Gillies and Joseph Connell, Metro-North commuter railroad (pers. comm.).
5. London et al. (1991) reported a statistically significant association between leukemia risk and regions using Δ-system primary lines (the Denver Wertheimer-Leeper wiring configuration) but found no effect in areas using neutral-wire Y systems.
6. The Biot-Savart law follows from Maxwell's equations in the static limit if we assume $\mathbf{B} = \mu\mathbf{H}$, where $\mathbf{B} = \nabla \times \mathbf{A}$ and solve $\nabla^2 \mathbf{A} = -\mu\mathbf{J}$. See Stratton (1941), 230–32. It should be understood that the integral of the differential form in eq. (24) should be taken over only one closed loop about the wire. Also see Bleil (1957).
7. See, e.g., Bartberger (1950).
8. See Bennett (1948).
9. See, e.g., Builder et al. (1953), chap. 5.
10. See, e.g., Rhoads (1987) and Finch et al. (1988).
11. See, e.g., Bennett (1976), chap. 7.
12. This particular method (which is original with the present text) works because the meter finds the time average of the square of the total field in each spatial direction separately in order to determine the rms values for each spatial component. In this process, all the cross terms between different frequency components (which generate sum and difference frequencies between the original frequencies present) average out to zero.
13. See, e.g., Bennett (1976), chap. 7.
14. See Bagule et al. (1975), 30–14.
15. See, e.g., Andrews (1993).
16. See also Jeans (1915), chap. 8.

17. To this end, Stratton (1941), 50–59, discusses the solution of Laplace's equation in cylindrical, spherical, elliptic, parabolic, bipolar, spheroidal, paraboloidal, and ellipsoidal coordinate systems.

18. The most complete account of the conjugate method appears to be the one given by Jeans (1915), chap. 8. Most subsequent texts merely repeat a few of the examples discussed by Jeans.

19. Unless otherwise indicated, "head level" will refer in this book to approximately 2 m above the ground.

20. See, e.g., Harnwell (1938).

21. The value of Q in this case was checked numerically to within 0.1 percent by integrating the vertical component of the field to ground.

22. See the discussion by Gönen (1988), 65, 66, 508.

23. Exact numerical calculations for the field near ground level from this geometry show that the errors introduced by approximation (eq. [39]) are only on the order of a few percent.

24. Because Δ systems usually incorporate lower voltages (e.g., 480-V, 2.4-kV, and 4.16-kV phase-to-phase rms voltages) in contemporary distribution power lines, they are not likely to be the largest sources of electric fields in most urban communities.

25. Gary Goedde, Cooper Power Systems, Fransville, Wisconsin (pers. comm.).

3
Natural Sources of Exposure

1. See Nagata (1974).

2. For a treatment of the electrodynamics of the trapping field, see Bennett (1976), chap. 5.

3. Because the magnetic field is a vector, rotation in the presence of even a uniform field results in a rate of change of magnetic induction. By turning your head back and forth through an angle of 90 degrees per second in the presence of the earth's field, you would experience a value of $|d\mathbf{B}/dt| \approx 1,400$ mG/s, a value that corresponds to an rms ambient field of about 2.6 mG at 60 Hz. (Of course, this process could introduce brain injury by itself.)

4. Merely as an example and not as an endorsement, the Magnum 310, made by the Dexsil Corporation of Hamden, Connecticut, provides simultaneous measurement of three-orthogonal field components, as well as the value of the resultant field of both the fundamental and third harmonics of 60-Hz fields over the range from about 0.04 mG to 17 G.

5. See Standard Procedures for Measurement of Power Frequency of Electric and Magnetic Fields from AC Power Lines, IEEE Standard 644–1987.

4
The Coupling of Electromagnetic Fields to the Body

1. Also see, e.g., Pethig (1988), E-62.

2. See, e.g., Page and Adams (1945), 89–113.

3. A derivation of the potential is given in Jeans (1915), 244–54, and is reproduced using the present notation in Stratton (1941), 207–11. But neither of these authors calculated the electrical field.

4. See Jahnke and Emde (1933), chap. 15.

5. Because the integrand is monotonically decreasing with increasing ξ, numerical integration can easily be accomplished to less than 0.1 percent error in a few seconds using the Trapezoidal Rule in a loop, where $d\xi' = kd\xi$, with $k \approx 1.001$, $d\xi_0 \approx 1E - 6$, and a convergence requirement $\approx 1E - 7$. (For example, the case $a = b = c = 1$ took fewer than 5 seconds for a compiled program on a Macintosh IIci.)

6. See Schwan (1983; 1988, 257).

5
Natural Sources of Noise

1. See Bennett (1960), 101–09.

2. See, e.g., Bennett (1960).

3. See Netter et al. (1978), 48.

4. See, e.g., Tolman (1946), 93.

5. For a derivation, see Bennett (1960), chap. 2.

6. Adair (1991) estimates that E_{kT} is about three thousand times larger than the internal fields produced by direct coupling from 300-V/m, 60-Hz fields in air. But because of distortion by the coupling process, the effective worst-case external fields from distribution lines can be much larger than 300 V/m.

7. See Hodgkin et al. (1952), table 1, or Hodgkin and Huxley (1952), table 3. Experimental values range from ≈ 0.8 to 1.5 $\mu F/cm^2$.

6
Observed Interactions with Electromagnetic Fields

1. See the review by Frankel (1986), 177–85.

2. See the literature review by Brady and Reiter (1992), VII-18–VII-31.

3. See the review by Brady and Reiter (1992).

4. See also Barnes (1986b), 122–28.

5. The tests were administered by the Rehabilitation Center of Eastern Fairfield County, Fairfield, Connecticut, an organization certified by the American Speech and Hearing Association for Speech Pathology and Audiology, according to Franklyn Beebe, president of Cortronix of Newton, Connecticut, who now manufactures the hearing aid (pers. comm.). (Schafer died in 1987.)

6. I once listened to an entire movement of a Beethoven string quartet over Schafer's apparatus. The fidelity was poor (considerable distortion and poor bass response), but the signal definitely was not swamped by noise and was not conducted through the normal hearing channels.

7. See the review by Postow and Swicord (1986), 425–60; esp. 443 ff.

8. See ibid., sec. 3.

9. See Lawrence and Cooksey (1936).

10. See Hodgkin et al. (1952), table 1, or Hodgkin and Huxley (1952), table 3. Experimental values range from ≈ 0.8 to 1.5 $\mu F/cm^2$.

11. When two operators commute, the eigen functions of one are eigen functions of the other. Hence, if a particular perturbation has an operator that commutes with the original Hamiltonian, the new eigen functions of that operator are also eigen functions of the original Hamiltonian (or, at worst, linear combinations thereof).

12. The oscillator would still have cylindrical symmetry about the z-axis. The Hamiltonian still commutes with L_z, but no longer with L^2. The problem becomes more complex but may be solved in cylindrical coordinates. See Bohm (1951), 351–60.

13. The electric-dipole transition ($\Delta l = \pm 1$) is normally the most probable "allowed transition" in atomic physics and represents the first term in a matrix element between the initial and final states for the operator $\exp(i2\pi r/\lambda)$, where r is the position of the charge from the origin, and λ is the transition wavelength. The next term (giving rise to electric-quadrupole transitions, with $\Delta l = 0, \pm 2$) is normally classified "forbidden" and is roughly lower in probability than the dipole approximation by the ratio $(2\pi r/\lambda)^2$. For atomic dimensions and optical wavelengths, $2\pi r/\lambda \approx 10^{-3}$ and the "forbidden lines" are down in probability by about four million. For $r \approx 1$ μm and $\lambda \approx 45$ μm (wavelength corresponding to $hf = kT$ at body temperature), as in the present case, the forbidden lines might only be down by about eight hundred.

14. See, e.g., Schiff (1949), 255.

15. Although the normalization coefficients and radial electric-dipole matrix elements are not given in the standard textbooks, the integrals involved are easily reduced to the one for the probability integral by noting that $\int r^{2n} \exp(-\beta r^2)\, dr = (-1)^n$ $\partial^n/\partial\beta^n \int \exp(-\beta r^2)\, dr$, where $n = 1, 2$.

16. These transitions were interpreted by Bowen in 1927 to be radiation on forbidden lines from very-long-lived states of oxygen and nitrogen that could not be observed in the laboratory. Because the transitions did not correspond to known transitions and came from a distant nebula, they were first identified as lines from a new element called "nebulium." As energy levels for oxygen and nitrogen were established more precisely, the forbidden-transition wavelengths could be calculated with enough precision to permit identification.

17. The "cross section" (σ) for a collision is the effective area of interaction swept out by one particle as it moves through an average density (N) of other particles in the medium. The total collision rate per particle is given by $N\sigma v$, where v is the relative velocity in the collision. A cross section of $\sigma = 10^{-14}$ cm^2 means that the colliding particle effectively has a cross-sectional area about one hundred times larger than it would in a billiard-ball model of the atom based on ground-state atomic dimensions.

18. See, e.g., Bennett (1962).

19. We can estimate the three-body collision rate by calculating the number of two-body collisions that occur during the time of one collision. The duration of one two-body collision is roughly $\sqrt{\sigma}/v$, where v is the relative velocity and σ is the cross section. We then multiply that number by the two-body collision rate per particle. Thus, the total three-body collision rate per particle would be about $N^2\sigma^{5/2}v \approx$ 300 MHz at one atmosphere for $\sigma = 10^{-14}$.

20. Spectral broadening from inelastic two-body collisions arises from a state that decays with time as $\exp(-Rt)$ where the total average decay rate is given by $R = A + N\sigma V$. Here, A is the total radiative decay rate, N is the density of ground state atoms, V is the relative velocity and σ is the inelastic destructive collision cross section. Fourier analysis of the exponential decay yields a full line width at half maximum energy of $R/2\pi$, $\approx N\sigma V/2\pi$, in the limit that A is negligible. (See, e.g., Bennett [1977], 95, 150.)

21. It may be shown from the Principle of Detailed Balancing that the combination of inelastic de-excitation and its inverse process lead precisely to populations for the nth excited level given by the Boltzmann factor, $g_n \exp(-E_n/kT)$, where g_n is the level degeneracy of energy level E_n.

Glossary

A   Ampere. A unit of current (see table 1.4)

Admittance   The equivalent of conductance for an AC circuit; the reciprocal of impedance

Amplitude   Magnitude

amu   Atomic mass unit (see table 1.3)

Anode   A positively charged electrode

Atomic transition   A change in state of an atom that results by absorption or emission of one quantum of energy (photon) through interaction with the electromagnetic field. For example, the emission or absorption of light.

Bandwidth   Range of frequencies over which an electric circuit or device responds. For resonant circuits, the full width at half maximum-energy response is roughly equal to the dissipative decay rate divided by 2π.

Bifilar winding   A winding made from wire that is wound back on itself so as to cancel external magnetic fields produced by current in the winding

Biot-Savart law   Law by which magnetic fields may be calculated from currents in wires (see eq. [24] and Chapter 2, note 6)

Boltzmann's constant   Mean thermal energy per degree of absolute temperature (see table 1.3)

Calcium-ion efflux   Movement of calcium ions across a cell membrane

Capacitance   Ratio of the change in charge of a conductor to its corresponding change in potential; also, a device for storing charge consisting of a pair of metal plates separated by a dielectric medium

Carrier frequency   The frequency of a radio wave used to transmit information

Charge   A quantity of electricity, measured in coulombs. Charge can be either negative (as with electrons) or positive (as with protons, or atomic nuclei). When there are equal amounts of positive and negative charge, the net charge is said to be neutral or zero.

Charge distribution   Continuous variation of charge over a volume or surface

CIRRPC   Committee on Interagency Radiation Research and Policy Coordination, which reports to CLSH

Closed-form solution   A solution to a problem given in terms of known analytic functions as opposed to a computed, numerical solution

CLSH   Committee on Life Sciences and Health, which reports to the Presidential Science Adviser

Coaxial cable   Cable made with concentric conductors separated by a dielectric. Such cable has broad frequency response and is shielded from pickup of external voltages.

Commutation   A mathematical operation in quantum mechanics. When two operators commute, the eigen functions of one are eigen functions of the other and the eigen values of each are "observables," or constants of the motion.

Conductance   Reciprocal of resistance. The current through a wire is equal to its conductance times the voltage applied across the wire.

Conductivity   Constant of proportionality in the equation relating the current density to the electric field in a conducting medium. The higher the conductivity, the greater the current density. Conductivity is the reciprocal of resistivity. (See tables 1.4 and 4.2.)

Confounding factor   Unknown factor that could be present in an epidemiological study that is associated with the suspected cause and that might be the actual cause of the disease

Conjugate potentials   Method of solving two-dimensional electrostatic potential problems by use of complex variables in Laplace's equation

Control group   In epidemiology, a reference group of people in similar circumstances who have not been exposed to a suspected cause of a particular disease

Corona discharge   Ionization of air around a conductor at high potential

Coulomb's law   The basic law of force between charged particles. Like charges repel, opposite charges attract each other. The force between two charges has a magnitude proportional to the product of the charges divided by the inverse square of the distance.

Curl ($\nabla \times$)   A vector differential operator that can be used to calculate the magnetic induction field from the vector potential

Degeneracy   The number of levels with the same energy in a quantum mechanical system

Δ (Delta) System   Three-phase power-transmission system that requires only three transmission wires. Although the three-source EMF's are of equal magnitude with respect to a common ground and shifted successively in phase by $120°$, a ground wire is not included in the transmission line. (See figure 2.8.)

Dielectric   An insulating medium containing bound charges that can be polarized (or displaced) in the direction of an applied electric field

Dielectric constant   Relative permittivity

Diffraction fringes   Interference maxima in diffraction patterns

Diffraction pattern   The pattern created by the constructive and destructive interference of light or other electromagnetic waves

Displacement   The electric displacement as defined by Maxwell is a field inside a dielectric modified from the electric field by the polarizability of the medium. It is proportional to the electric field through a constant known as the permittivity. (See tables 1.2 and 1.4.)

Distribution line   Intermediate-voltage (typically, 12 kV) power line used on urban streets to deliver power to step-down transformers that supply 120/240 volt service to homes

Divergence ($\nabla \cdot$)   A differential operator that permits calculating the total differential change of a vector along three orthogonal components

Eigen function   A solution to an equation of the type $H\psi = E\psi$ (the time-independent Schrödinger equation), in which H is a differential operator (here, the Hamiltonian), E is a constant or "eigen value" (here, the energy) and ψ is the eigen function (or wave function). These functions were initially studied by English physicists and mathematicians, who called them "proper functions" and labeled the constants "proper values." When Schrödinger published his equation, he translated *proper value* as *eigenwert*. When English physicists read his articles they retranslated only part of each term; hence, the hybrid terms *eigen value*, *eigen function*, and *eigen value equation*.

Eigen value   *See* eigen function

Electrical noise   Spurious fluctuating voltages, currents, or fields

Electric dipole   Two equal- and opposite-point electric charges placed a small distance apart. The electric-dipole moment is the product of the common charge magnitude and the charge separation. In quantum mechanics, an electric-dipole transition is the dominant type of "allowed transition." The transition probability depends on matrix elements of the electric-dipole-moment operator. (See eq. [92] and related discussion.)

Electric field   Force per unit electric charge produced by distributions of other electric charges

ELF   Extremely low frequency, defined by international convention as within the band from 30 to 300 Hz

EMF   Electromotive force. Voltage drop across a battery or generator terminals. Unfortunately, the term has been introduced by the popular press to stand for *electromagnetic field* or *electric and magnetic fields*. It is not so used in this book.

Equipotential   Surface of constant potential

EPRI   Electric Power Research Institute (Palo Alto, California)

eV   Electron volt. The energy acquired by an electron in falling through a potential of 1 V (see table 1.3).

F   Farad. (An enormous) unit of capacitance, named after Michael Faraday (see table 1.4)

Faraday cage   An enclosure of grounded conducting material, like copper screening, that provides extremely high shielding from external electric fields because of the properties of Maxwell's equations

Faraday's law   The electric field produced in a closed loop of wire is proportional to the time rate of change of the total magnetic flux enclosed by the wire (see eq. [66])

Far field   A field at a distance from the source that is large compared with the wavelength

Field   Force per unit charge acting at a distance from the source. Fields are vector quantities.

Fourier analysis   A mathematical method for determining harmonic (or spectral) content of a periodic waveform that is named after its inventor, Baron Joseph von Fourier

Free-space values   Values that are measured in vacuum (such as the velocity and wavelength of a light at a given frequency) as opposed to the values that are measured in matter

Frequency   The rate at which a periodic waveform repeats itself in time at one position in space. Frequency is in units of cycles per second (Hz) (see table 1.1).

G   Gauss. A unit of magnetic flux in cgs (centimeter-gram-second) system

Gauss's theorem   The integral of the outward normal component of a vector over a surface area enclosing a bounded volume is equal to the integral of the divergence of that vector over the volume. When used with the second of Maxwell's equations (see table 1.2), it provides a useful way to calculate displacement fields in terms of the charge within the surface.

Gradient (∇)   A vector differential operator which may be used to evaluate the electric field from a scalar potential

Ground   An electrical ground is a large conductor at a neutral or zero voltage in respect to other components in the system. It might be the earth itself, or a metal chassis on which an electric circuit is constructed.

Ground loop   An extra conduction path between electrical grounds (through a water pipe, for example) that gives rise to an unexpected current loop

Hamiltonian   The total energy operator in the Schrödinger equation based on a corresponding quantity defined in classical mechanics by Hamilton

Harmonic   The integral multiple of the fundamental frequency of a periodic wave form

Harmonic analysis   *See* Fourier analysis

Helmholtz coils   A pair of parallel large-diameter coils designed to produce a nearly uniform magnetic field near their common midpoint

High-tension lines   Very high voltage (typically, 100–500 kV) transmission lines used to deliver power over long distances

Hz   Hertz. A unit of frequency, named in honor of Heinrich Hertz, who discovered radio waves. 1 Hz = 1 cycle per second (see table 1.1).

Image pair   A pair of imaginary charges used to solve electric field problems in the method of images

Impedance   The equivalent of resistance for an AC circuit

Inductance   A coil of wire, often wound on a core of high permeability metal such as iron

Intensity   As applied to fields, the square of the field amplitude. In most classical (nonquantum) phenomena, the flow of energy or power is proportional to the amplitude squared, hence, to the intensity.

Ionizing radiation   Electromagnetic radiation for which the frequency is high enough so that the energy per quantum exceeds the binding energy of electrons in atoms and molecules. Such frequencies are typically in the optical range or higher, hence more than a trillion times larger than power-line frequencies.

J   Joule. A unit of energy (see table 1.4)

Johnson noise   Thermal noise induced in a resistor, which was discovered experimentally by J. B. Johnson and first explained theoretically by Harry Nyquist (see eq. [69])

kT   The mean thermal energy at absolute temperature, T; k is Boltzmann's constant (see table 1.3)

Laplace's equation   The differential equation for the electric (scalar) potential in a charge-free region. The potential determined from the equation can be used to compute the electric field in such a region (see eq. [32]).

Linear   A proportional relation. A linear resistance, for example, is one across which the voltage decrease is proportional to the current flowing through it and for which a graph of the voltage versus the current would be a straight line.

Line voltage   Electric potential difference between pairs of wires on the line (see discussion of eq. [23])

Load   The resistance or impedance placed across a line

Lorentz force   The force acting on a charged particle moving through combined electric and magnetic fields (see eq. [8])

Magnetic dipole   A magnetic field from a current loop that falls off as the cube of the distance at large distances (see eq. [27]). The far field from such a current loop is equivalent to the field obtained from two opposite hypothetical-unit magnetic poles a small distance apart.

Magnetic flux   The product of the magnetic induction field with a differential element of area normal to the field (see table 1.4)

Magnetic shield   Any high-permeability material (soft iron, for instance) that confines magnetic flux to a defined path

Matrix element   In quantum mechanics, an integral of the type $\int \psi_m^* \, O \, \psi_n \, dV$, where O is an operator and ψ_m and ψ_n are eigen functions of the Hamiltonian

Maxwell's equations   A set of unified equations that relate the electric and magnetic fields with the charge and current densities producing them (see table 1.2)

Mean value   The sum of the number of events times the probability of the event over a normalized probability distribution

Melatonin   A molecule secreted from the pineal gland in the brain

Method of images   A method for solving electrostatic-field problems by placing imaginary point charges of opposite sign so as to yield equipotential surfaces equivalent to those represented by conductors in a given situation

Microwave radiation   Electromagnetic radiation with wavelengths in the range from about 1 mm to 30 cm — hence, frequencies ranging from about 1 to 300 billion cycles per second (see figure 1.1)

Molecular transition   A transition similar to an atomic transition, except that it pertains to a molecule

Mu metal   A thin, high-permeability metal that can be used to produce shielding against small magnetic fields

Near field   A field at a distance from the source that is small compared to the wavelength

Nonlinear   A nonproportional relation. A nonlinear resistance, for example, is one across which the voltage decrease is not simply proportional to the first power of the current flowing through it.

Normal distribution   Gaussian distribution

Normalized   A normalized probability distribution has been multiplied by a constant such that the sum of the distribution over all possible outcomes is unity

Oblate spheroid   A spheroid that is shortened along the axis of rotational symmetry

Ω (Ohm)   A unit of resistance, named after George Ohm, who discovered Ohm's law (the voltage decrease across a resistance equals the current times the resistance)

Ohmic heating   Power dissipation by resistive loss. This loss equals the resistance times the square of the current flowing through it.

ORAU   Oak Ridge Associated Universities. A private nonprofit consortium of universities with a mission to provide capabilities critical to the nation's technology infrastructure (Washington, D.C., and Oak Ridge, Tennessee).

Period   The time interval over which a periodic waveform repeats itself; the reciprocal of the fundamental frequency (see table 1.1)

Periodic waveform   A waveform that repeats itself precisely

Permeability   A quantity that measures a material's influence on magnetic flux; the constant of proportionality in the equation relating the magnetic induction to the magnetic field in a medium (see tables 1.3 and 1.4)

Permittivity   A quantity that measures a material's ability to store electrical energy in an electric field; the constant of proportionality in the equation relating the electric displacement to the electric field in a medium (see tables 1.3, 1.4, and 4.1)

Phase   The angular displacement of a sinusoidal waveform; effectively a shift in time at a given frequency

Pineal gland   A small, pinecone-shaped gland behind the third ventricle of the brain

Planck's constant (h)   The constant of proportionality in the energy per quanta as a function of frequency (see table 1.3)

Poisson's equation   A modification of Laplace's equation for the electric potential in order to include the presence of charge distributions

Positive correlation   In epidemiology, causative association

Potential   Work done per unit charge in moving against an electric field (see table 1.4)

Power   The rate of flow of energy. In an electromagnetic wave, the power is proportional to the intensity of the wave, or the amplitude squared (more precisely, the product of the electric- and magnetic-field amplitudes).

Poynting vector   A quantity that indicates the power flux and direction of radiation in an electromagnetic wave. It is proportional to the product of the electric and magnetic fields and moves in a direction normal to the plane formed by those two field vectors.

Prolate spheroid   A spheroid that is elongated along the axis of rotational symmetry

Q   Quality factor. A measure of the sharpness of a resonance, formally defined as 2π times the energy stored divided by the energy lost per cycle. For most resonant systems, Q is approximately equal to the resonant response frequency divided by the bandwidth (full width at half-maximum energy response).

Quantum   The smallest unit of energy that can be exchanged with an electromagnetic wave. The quantum size is proportional to the frequency of the wave through Planck's constant (see table 1.3).

Quantum theory   The theory of quantized atomic energy levels and their interaction with light

Random process   Something that occurs with equal likelihood for all possible outcomes

Relative permittivity   The ratio of the permittivity for a particular material to the permittivity for free space; the dielectric constant

Relativistic mass   The increased effective mass obtained when a particle moves at close to the speed of light (see discussion after eq. [9])

Resistance   The constant of proportionality in Ohm's law. The voltage decrease along a wire is equal to the current times the resistance. The unit of resistance is the ohm (Ω).

Resistivity   The resistance of a bulk material measured in ohm-meters (Ωm). Resistivity is the reciprocal of conductivity.

Resonant circuit   An electric circuit that responds selectively within a narrow range about some frequency of peak response; for example, a capacitance and an inductance connected in parallel

Rest mass   The mass of a particle measured in a reference frame in which it is not moving

Right-of-way   As applied to the power industry, a strip of land over an individual's ground that is set aside by usage or formal agreement for power lines

Sawtooth waveform   A periodic waveform that looks like the profile of a rip saw when displayed in time

Scalar   A quantity, like the electric potential, that possesses magnitude but not direction

Scalar potential   A potential from which the electric field may be calculated

Schrödinger equation   The basic partial-differential equation defining the wave nature of atoms and molecules. *See also* eigen function.

Shot noise   Fluctuations in current owing to the discreteness of electrical charge

Sidebands   Frequencies spaced about the carrier frequency of a radio wave that are generated by modulation

Single-phase line   A power line in which all voltages or currents have the same phase angle

Sinusoid   A waveform based on the shape of the sine function

Spheroid   A spherelike object that has an axis of rotational symmetry but that is shortened (oblate) or elongated (prolate) along that axis

Standard deviation   In statistics, the square root of the variance. Results normally are expected to fall within one standard deviation of the mean 67 percent of the time; within two standard deviations 95 percent of the time, and so on.

Steady state   An equilibrium solution, or unvarying condition, that prevails after the transients associated with turning a voltage or current on (or off) have died out

Stokes's law   The resistive force on a sphere of radius a moving slowly at a velocity v through a liquid given by $6\pi\eta a v$, where η is the viscosity of the liquid

Stokes's theorem   The line integral of a vector around a closed loop is equal to the integral of the outward normal of the curl of that vector over the surface area enclosed by the loop. The theorem is particularly useful in calculating the magnetic-induction fields produced by currents flowing in loops of wire.

T Absolute temperature in degrees Kelvin (see table 1.3)

T Tesla. A unit of magnetic flux (see table 1.4)

Thermal fields   Electric fields produced by the random motion of electrons or ions in resistive material. They increase with the square root of the absolute temperature.

Thermal noise   Fluctuating voltage produced by random motion of electrons or ions in resistive material that increases with the square root of the absolute temperature; also known as "Johnson noise"

Three-phase line   A power line requiring a minimum of three separate wires in which the voltages and currents differ successively by a phase angle of $120°$ (see figure 2.8)

Tomography   A method used in radiography to display cross sections within the body

Torque   Moment of force causing a body to rotate; product of the lever arm about an axis of rotation and the force acting perpendicularly to both the lever arm and the axis

Transformer   Two (or more) coils insulated from each other but wound on a common (often high-permeability) core. Alternating magnetic flux produced by the current in one coil induces a voltage from the Faraday effect across the other coil. By varying the relative number of turns in the two coils, the voltage across the first coil can be stepped down or stepped up across the second coil. When connected to a load resistance, an increase in voltage in the secondary winding is accompanied by a proportionate decrease in current, and vice versa.

Transmission lines   As distinct from distribution lines, wires carrying electricity over large distances, usually at very high voltage (hence, lower currents) in order to minimize power loss. Three-phase transmission is generally used for this purpose.

Unified equations   A reduced set of equations that includes the mathematical description of a large number of seemingly diverse phenomena. For example, Maxwell's equations include all the basic laws of electricity and magnetism that before Maxwell had required about a dozen separate equations (Coulomb's law, Faraday's law, Henry's law, Gauss's theorem, the law of Biot-Savart, Laplace's equation, Poisson's equation, and so on).

V Volt. A unit of electric potential (see table 1.4)

Van Allen belts   Regions in which high-energy protons from the sun are trapped by the earth's magnetic field

Variance   The average over a probability distribution of the square of the difference from the mean

Vector potential   A vector quantity from which the magnetic induction field may be calculated

VLF   Very low frequency, defined by international convention to occupy the band from 3 to 30 kHz

Waveform   A curve showing the shape of a wave as a function of time

Wavelength   The distance over which a periodic wave repeats itself at one instant in time. For electromagnetic waves, the wavelength equals the velocity of light divided by the frequency.

Windows   Some scientists have reported narrow regions ("windows") of field amplitude and/or frequency that interfere with calcium-ion efflux through cell membranes. These reports are controversial.

Y system   A three-phase power transmission or distribution system (sometimes called a "star system") that requires four wires, one of them being the common electrical ground. As with the delta system, the three-source EMFs are of equal magnitude with respect to the common ground and shift successively in phase by 120° (see figure 2.8).

Bibliography

Aaron, R. K.; Ciombor, D. McK.; Jolly, G. 1989. Stimulation of experimental endochondral ossification by low-energy pulsing electromagnetic fields. J. Bone Miner. Res. 4(2):227–33.

Adair, E. R. 1991. "Currents of Death" rectified: A paper commissioned by the IEEE-USA Committee on Man and Radiation in response to the book by Paul Brodeur. New York: IEEE-USA.

Adair, R. K. 1991. Constraints on biological effects of weak extremely low-frequency electromagnetic fields. Phys. Rev. A43:1039–48.

———. 1992. Criticism of Lednev's mechanism for the influence of weak magnetic fields on biological systems. Bioelectromagnetics 13:231–35.

———. 1993. Effects of ELF magnetic fields on biological magnetite. Bioelectromagnetics 14:1–4.

Adey, W. R. 1980. Frequency and power windowing in tissue interactions with weak electromagnetic fields. Proc. IEEE 68:119–25.

Akasofu, S.-I. 1974. Encyclopaedia Britannica, 15th ed., s.v. "Auroras."

Altman, L. K. 1993. At AIDS talks, science confronts daunting maze. New York Times International, 6 June, 20.

Andrews, E. L. 1993. Top rivals agree on unified system for advanced TV. New York Times, 25 May.

Bagule, E.; et al., eds. 1975. Reference data for radio engineers. New York: Howard Sams.

Barnes, F. S. 1986a. Interaction of DC electric fields with living matter. In: Polk, C.; Postow, E., eds., CRC handbook of biological effects of electromagnetic fields. Boca Raton, Fla.: Chemical Rubber Company Press, 99–119.

———. 1986b. Extremely low frequency (ELF) and very low frequency electric fields: Rectification, frequency sensitivity, noise, and related phenomena. In: Polk, C.; Postow, E., eds., CRC handbook of biological effects of electromagnetic fields. Boca Raton, Fla.: Chemical Rubber Company Press, 121–38.

Barnes, F. S.; Hu, C. L. 1977. Model for some non-thermal effects of radio and microwave fields on biological membranes. IEEE Trans. Microwave Technol. 25:742.

Barnes, H. C.; McElroy, A. J.; Charkow, J. H. 1967. Rational analysis of electric fields in live line working. IEEE Trans. Power App. Syst. 8:482–92.

Bartberger, C. L. 1950. The magnetic field of a plain circular loop. J. Appl. Phys. 21:1108.

Bassett, C. A. L. 1968. Biologic significance of piezoelectricity. Calcif. Tissue Res. 1:252–72.

Bassett, C. A. L.; Calo, N.; Kort, J. 1981. Congenital "pseudarthroses" of the tibia: Treatment with pulsing electromagnetic fields. Clin. Orthop. Rel. Res. 154, sec. 2:136–49.

Bassett, C. A. L.; Pilla, A. A.; Pawluk, R. J. 1977. A nonoperative salvage of surgically resistant pseudarthroses and nonunions by pulsing electromagnetic fields: A preliminary report. Clin. Orthop. Rel. Res. 124:128–42.

Baumeister, T.; Marks, L. S., eds. 1967. Standard handbook for mechanical engineers. New York: McGraw-Hill.

Bawin, S. M., Adey, W. R. 1976. Sensitivity of calcium binding in cerebral tissue to weak environmental electric fields oscillating at low frequencies. Proc. Natl. Acad. Sci. (USA) 73:1999–2003.

Bawin, S. M.; Gavalas-Medici, R. J.; Adey, W. R. 1973. Effect of modulated very high frequency fields on specific brain rhythms in cats. Brain Res. 58:365.

Bawin, S. M.; Kaczmarek, K. L.; Adey, W. R. 1975. Effects of modulated VHF fields on the central nervous system. Ann. N.Y. Acad. Sci. 247:74–91.

Becker, R. O. 1978. Electrical osteogenesis: Pro and con. Calcif. Tissue Res. 26:93–97.

Bennett, M. V. L.; Spira, M. E.; Pappas, G. O. 1972. Properties of electronic functions between embryonic cells of *Fundulus*. Dev. Biol. 29:419–35.

Bennett, W. R. 1948. The spectra of quantized signals. Bell Syst. Tech. J. 27:446–72.

———. 1960. Electrical noise. New York: McGraw-Hill.

Bennett, W. R., Jr. 1962. Optical spectra excited in high pressure noble gases by alpha impact. Ann. Physics 18:367–420.

———. 1976. Scientific and engineering problem solving with the computer. Englewood Cliffs, N.J.: Prentice-Hall.

———. 1977. The physics of gas lasers. New York: Gordon and Breach.

———. 1992. General ellipsoidal model for coupling electric fields to the human body. In: Health effects of low-frequency electromagnetic fields. Oak Ridge, Tenn.: ORAU, pp. II-63–II-69.

Blackman, C. F.; Benane, S. G.; Elliott, D. J.; House, D. E.; Pollock, M. M. 1988. Influence of electromagnetic fields on the efflux of calcium ions from brain tissue *in vitro*: A three-model analysis consistent with the frequency response up to 510 Hz. Bioelectromagnetics 9(3):215–27.

Blackman, C. F.; Benane, S. G.; House, D. E.; Elliott, D. J. 1990. Importance of alignment between local DC magnetic field and an oscillating magnetic field in responses of brain tissue *in vitro* and *in vivo*. Bioelectromagnetics 11:159–67.

Blackman, C. F.; Benane, S. G.; House, D. E.; Joines, W. T. 1985. Effects of ELF (1–120 Hz) and modulated (60 Hz) RF fields on the efflux of calcium ions from brain tissue *in vitro*. Bioelectromagnetics 6(1): 1–11.

Blackman, C. F.; Benane, S. G.; Kinney, L. S.; Joines, W. T.; House, D. E. 1982. Effects of ELF on calcium-ion efflux from brain tissue *in vitro*. Radiat. Res. 92(3): 510–20.

Blackman, C. F.; Benane, S. G.; Rabinowitz, J. R.; House, D. E.; Joines, W. T. 1985. A role for the magnetic field in the radiation-induced efflux of calcium ions from brain tissue *in vitro*. Bioelectromagnetics 6(4):327–37.

Blackman, C. F.; Elder, J. A.; Weil, C. M.; Benane, S. G.; Eichinger, D. C.; House, D. E. 1979. Induction of calcium-ion efflux from brain tissue by radio frequency radiation: Effects of modulation frequency and field strength. Radio Sci. 14:93–98.

Bleil, D. F., ed. 1957. Electricity and magnetism. American Institute of Physics Handbook. New York: McGraw-Hill, sect. 5.

Bohm, D. 1951. Quantum theory. New York: Prentice Hall.

Bonneville Power Administration. June 1989. Electrical and biological effects of transmission lines. Portland, Oregon. DOE/BPA-961, p. 14.

Book, D. L. 1987. NRL plasma formulary. Washington, D.C.: Naval Research Laboratory.

Bowen, I. S. 1927. The origin of the nebulium spectrum. Nature 120:473.

Brady, J. V.; Reiter, R. J. 1992. The pineal gland. In: Health effects of low-frequency electromagnetic fields. Oak Ridge, Tenn.: ORAU, VII-18–VII-31.

Brent, R. L.; Gordon, W. E.; Bennett, W. R.; Beckman, D. A. 1993. Reproductive and teratologic effects of electromagnetic fields. Reproductive toxicology 7: 535–80.

Bridgewater, T. H.; Fink, D. G. 1974. Encyclopaedia Britannica, 15th ed., s.v. "Television."

Brighton, C. T.; Black, J.; Friedenberg, Z. B.; Esterhai, J. L., Jr.; Day, L. J.; Connolly, J. F. 1981. A multicenter study of the treatment of non-union with constant direct current. J. Bone Joint Surg. Am. 63:2–42.

Brodeur, P. 1989. Annals of radiation: The hazards of electromagnetic fields. Parts 1–3. New Yorker, 12 June, 51–88; 19 June, 47–73; 26 June, 39–68. (Later published as Currents of death: Power lines, computer terminals and the attempt to cover up their threat to your health. New York: Simon and Schuster, 1989.)

————. 1992. Annals of radiation: The cancer at Slater School. New Yorker, 7 Dec., 86–89.

Builder, G.; Hansen, I. C.; Langford-Smith, F. 1953. Transformers and iron-cored inductors. In: Langford-Smith, F., ed., Radiotron designer's handbook. Harrison, N.J.: RCA.

Cramér, H. 1946. Mathematical methods of statistics. Princeton, N.J.: Princeton University Press.

Christel, P.; Ceff, G.; Pilla, A. A. 1981. Modulation of rat radial osteotomy repair using electromagnetic current induction. In: Becker, R. O., ed., Mechanisms of growth and control. Springfield, Ill.: C. C. Thomas.

Cooper, M. S. 1984. Gap junctions increase the sensitivity of tissue cells to exogenous electric fields. J. Theor. Biol. 111:123–30.

Croft, T.; Summers, W. I. 1992. American electrician's handbook. New York: McGraw-Hill.

Davis, J. G.; et al. 1993. EMF and cancer. Science 260:13.

Davydov, A. S. 1982. Biology and quantum mechanics. Elmsford, N.Y.: Pergamon Press.

DeFelice, L. 1981. Introduction to membrane noise. New York: Plenum Press.

Dolezalek, H. 1988. Atmospheric electricity. In: Weast, R. C.; et al., eds., CRC handbook of chemistry and physics. 69th ed. Boca Raton, Fla.: Chemical Rubber Company Press, p. F-156.

Feychting, M.; Ahlbom, A. 1992. Magnetic fields and cancer in people residing near Swedish high voltage power lines. Stockholm: Karolinska Institute. IMM report 8/92.

Finch, S. R.; Lavigne, D. A.; Scott, R. P. W. 1988. One example where chromatography may not necessarily be the best analytical method. J. Chromatogr. Sci. 28:351–56.

Flannery, J. T.; Polednak, A. P.; Fine, J. A.; Mayne, S. T.; Mayne, J. T.; Benn, S. L.; Hampton, M. M.; Kegels, S. S.; Morra, M. E.; Warren, S. L. 1992. Cancer trends and cancer prevention research in Connecticut: A statistical and epidemiological report, 1935–1986. New Haven: State of Connecticut Department of Health: Cancer Prev. Research Unit for Connecticut at Yale, Connecticut Tumor Registry, Yale Comprehensive Cancer Center.

Floderus, B.; et al. 1992. Occupational exposure to electromagnetic fields in relation to leukemia and brain tumors: A case control study. Solna, Sweden: National Institute of Occupational Health.

Florig, H. K.; Nair, I.; Morgan, M. G. 1987. Briefing paper 1: Sources and dosimetry of power frequency fields. Technical report prepared for the Florida Dept. of Environmental Regulation. DER contract SP117.

Foster, K. R.; Finch, E. F. 1974. Microwave hearing: Evidence for thermoacoustic auditory stimulation by pulsed microwaves. Science 185(147): 256–58.

Foster, K. R.; Schwan, H. P. 1986. Dielectric properties of tissue. In: Polk, C.; Postow, E., eds., CRC handbook of biological effects of electromagnetic fields. Boca Raton, Fla.: Chemical Rubber Company Press, 27–119.

Frankel, R. B. 1986. Biological effects of static magnetic fields. In: Polk, C.; Postow, E., eds., CRC handbook of biological effects of electromagnetic fields. Boca Raton, Fla.: Chemical Rubber Company Press, 169–96.

Frey, A. H. 1961. Auditory system response to RF energy. Aerosp. Med. 32:1140.

———. 1962. Human auditory system response to modulated electromagnetic energy. J. Appl. Physiol. 17:689.

Fröhlich, H. 1968. Long-range coherence and energy storage in biological systems. Int. J. Quant. Chem. 2:64.

Gauger, A., Jr. 1985. Household appliance magnetic field survey. IEEE transactions on power apparatus and systems. PA-104 (Sept.)

Gönen, T. 1988. Modern high power systems. New York: Wiley.

Goodman, R.; Bassett, C. A. L.; Henderson, A. S. 1983. Pulsing electromagnetic fields induce cellular transcription. Science 220:1283–85.

Gordis, L.; Greenhouse, S. W. 1992. Relevant principles of epidemiology and data analysis. In: Health effects of low-frequency electric and magnetic fields. Oak Ridge, Tenn.: ORAU, chap. 3.

Harnwell, G. P. 1938. Principles of electricity and electromagnetism. New York: McGraw-Hill.

Hess, W. N.; Greisen, K. I. 1974. Encyclopaedia Britannica, 15th ed., s.v. "Van Allen radiation belts."

Hodgkin, A. L.; Huxley, A. F. 1952. A quantitative description of membrane current and its application to conduction and excitation in nerve. J. Physiol. 117:500–544.

Hodgkin, A. L.; Huxley, A. F.; Katz, B. 1952. Measurement of current-voltage relations in the membrane of the giant axon of Loligo. J. Physiol. 116:424–48.

Irabarne, J. V.; Cho, H. R. 1980. Atmospheric physics. Boston: Reidel.

Jackson, J. D. 1975. Classical electrodynamics. New York: Wiley.

————. 1992. Are the stray 60-Hz electromagnetic fields associated with the distribution and use of electric power a significant cause of cancer? Proc. Nat. Acad. Sci. (USA) 89:3508–10.

Jahnke, E.; Emde, F. 1933. Tables of functions. Leipzig: Teubner.

Jauchem, J. R. 1992. Epidemiologic studies of electric and magnetic fields and cancer: A case study of distortions by the media. J. Clinical Epidem. 45:1137–42.

Jeans, J. H. 1915. Electricity and magnetism. Cambridge: Cambridge University Press.

Johnson, J. B. 1928. Thermal agitation of electricity in conductors. Phys. Rev. 32:97–109.

Johnson, C. C.; Spitz, M. R. 1990. Childhood nervous system tumors: An assessment of risk associated with paternal occupations involving use, repair or manufacture of electrical and electronic equipment. Int. J. Epidem. 18:756–62.

Kaune, W. T.; Forsythe, W. C. 1985. Dosimetric comparison of human, rat, and pig models exposed to power frequency electric fields. Proc. of the US-USSR workshop on physical factors: Microwave and low frequency fields. Report of the National Institute of Environmental Health Sciences.

Kaune, W. T.; Phillips, R. D. 1980. Comparison of the coupling of grounded humans, swine and rats to vertical 60-Hz electric fields. Biomagnetics 1: 117.

Kavet, R. 1991. An alternative hypothesis for the association between electrical wiring configurations and cancer. Epidem. 2:224–29.

Kirschvink, J. L. 1981. Ferromagnetic crystals (magnetite?) in human tissue. J. Exp. Biol. 92:333–35.

Lawrence, E. O.; Cooksey, D. 1936. On the apparatus for the multiple acceleration of light ions to high speeds. Phys. Rev. 50:1131–44.

Lednev, V. V. 1991. Possible mechanism for the influence of weak magnetic fields on biological systems. Bioelectromagnetics 12:71–75.

Leonard, A.; Neutra, R.; Yost, M.; Lee, G. 1991. Suggested protocol for measuring 60 Hz magnetic fields in residences. Health Phys. Soc. Newsl. 19:13, 35–37 (Oct.).

Lerchl, A.; Nonaka, K. O.; Reiter, R. J. 1991. Pineal gland "magnetosensitivity" to static magnetic fields is a consequence of induced electric currents (eddy currents). J. Pineal Res. 10(3):109–16.

Lerchl, A.; Nonaka, K. O.; Stokkan, K. A.; Reiter, R. J. 1990. Marked rapid alterations in nocturnal pineal serotonin metabolism in mice and rats exposed to weak intermittent magnetic fields. Biochem. Biophys. Res. Comm. 169:102–08.

Liboff, A. R. 1985. Cyclotron resonance method for electromagnetic energy transfer to cells. Abs. Seventh Annual Mtg. Bioelectromagnetics Soc.

Liboff, A. R.; McCleod, B. R.; Smith, S. D. 1992. Method and apparatus for controlling plant growth. U.S. patent no. 5,077,934 (7 Jan.).

Liboff, A. R.; Rozek, R. J.; Sherman, M. L.; McLeod, B. R.; Smith, S. D. 1987. Ca^{2+}–45 cyclotron resonance in human lymphocytes. J. Bioelectricity 8:12–22.

Liboff, A. R.; Williams, T., Jr.; Strong, D. M.; Wistar, R., Jr. 1984. Time varying magnetic fields: Effect on DNA synthesis. Science 223:818–20.

Linscott, S. 1991. Meadow street cancer cluster. Health Phys. Soc. Newsl. 19:12.

Loewenstein, W. R. 1966. Permeability of membrane junctions. Annals N.Y. Acad. Sciences 137:441–72.

London, S. J.; Thomas, D. C.; Bowman, J. D.; Sobel, E.; Chen, T.-S.; Peters, J. M. 1991. Exposure to residential electric and magnetic fields and risk of childhood leukemia. Am. J. Epidem. 134(9): 923–37.

McLeod, B. R.; Smith, S. D.; Cooksey, K. E.; Liboff, A. R. 1987. Ion cyclotron resonance frequencies enhance Ca^{2+} mobility in diatoms. J. Bioelectricity 6:1–12.

McLeod, B. R.; Smith, S. D.; Liboff, A. R. 1987. Calcium and potassium cyclotron resonance curves and harmonics in diatoms (*A. coffaeformis*). J. Bioelectricity 6:153–68.

Matanoski, G. M.; Elliott, E. A.; Breysse, P. N. 1989. Cancer incidence in New York telephone workers. Poster presented at the annual review of research on biological effects of 50/60 Hz electric and magnetic fields, air ions, and ion currents, 15 November 1989, Portland Oregon. U.S. Department of Energy, Office of Energy Storage and Distribution.

———. 1991. Electromagnetic field exposure and male breast cancer. Lancet 337:737.

Maxwell, J. C. [1873] 1954. A treatise on electricity and magnetism. 2 vols. New York: Dover Publications.

Mills, D. W.; Rhoads, K. 1985. CLOR-N-OIL™ test kit as a PCB screening tool. Proc. 1985 EPRI PCB Sem., 22–25 Oct., pp. 4-7-4-14.

Morgan, M. G.; et al. 1989. In: Electric and magnetic fields from 60-hertz electric power. Carnegie Mellon University, Department of Engineering and Public Policy.

Nagata, T. 1974. Encyclopaedia Britannica, 15th ed., s.v. "Magnetic field of the earth."

Netter, F. N.; et al. 1978. Physiology and pathophysiology: The electrocardiogram. In: Yonkman, F. F., ed., The CIBA collection of medical illustrations, vol. 5: Heart. Summit, N.J.: CIBA.

Neyton, J.; Trautmann, A. 1985. Single-channel currents of an interstellar function. Nature 317:331–35.

Nicolet, M. 1974. Encyclopaedia Britannica, 15th ed., s.v. "Atmosphere."

Nyquist, H. 1928. Thermal agitation of electric charge in conductors. Phys. Rev. 32:110–13.

Orville, R. E. 1974. Encyclopaedia Britannica, 15th ed., s.v. "Lightning."

Page, L.; Adams, N. I., Jr. 1945. Principles of electricity. New York: Van Nostrand, 90–113.

Partridge, J. F. 1967. Railway engineering. In: Baumeister, T.; Marks, L. S., eds., Standard handbook for mechanical engineers. New York: McGraw-Hill.

Pethig, R. 1988. Relative permittivity and conductivity for various biological tissues at frequencies commonly used for therapeutic purposes. In: Weast, R. C.; et al., eds., CRC handbook of chemistry and physics. 69th ed. Boca Raton, Fla.: Chemical Rubber Company Press, p. E-62.

Pilla, A. A.; Nasser, P. R.; Kaufmann, J. J. 1992. The sensitivity of cells and tissues to weak electromagnetic fields. In: Cleary, A. J.; et al., eds., Charge and Field Effects in Biosystems-3. New York: Plenum Press, 231–41.

Plonsey, R. 1969. Bioelectric phenomena. New York: McGraw-Hill.

Polk, C. 1986. Introduction. In: Polk, C.; Postow, E., eds., CRC handbook of biological effects of electromagnetic fields. Boca Raton, Fla.: Chemical Rubber Company Press.

Poole, C.; Trichopoulos, D. 1991. Extremely low-frequency electric and magnetic fields and cancer. Cancer Causes Control 2:267–76.

Postow, E.; Swicord, M. L. 1986. Modulated fields and "window" effects. In: Polk C.; Postow, E., eds., CRC handbook of biological effects of electromagnetic fields. Boca Raton, Fla.: Chemical Rubber Company Press.

Reuss, S.; Olese, J.; Vollrath, I.; Skalej, M.; Meves, M. 1985. Lack of effect of NMR-strength magnetic fields on rat pineal melatonin synthesis. ICRS Med. Sci. 13:471.

Rhoads, K. W. 1987. CLOR-N-OIL™ test kit as a risk management tool: An update. Proc. 1987 EPRI PCB Sem., Dec.

Rice, S. O. 1944. Mathematical analysis of random noise. Bell Syst. Tech. J. 23:252–332.

———. 1945. Mathematical analysis of random noise. Bell Syst. Tech. J. 24:46–156.

Robinson, F. N. H. 1974. Encyclopaedia Britannica, 15th ed., s.v. "Electricity."

Rosch, W. L. 1990. Low-frequency electromagnetic fields: Unsafe at any frequency? MacUser (Feb.):147–51

Savitz, D. A.; Calle, E. E. 1987. Leukemia and occupational exposure to electromagnetic fields: Review of epidemiological surveys. J. Occ. Med. 29:47–51.

Savitz, D. A.; Wachtel, H.; Barnes, F. A.; John, E. M.; Turdik, J. G. 1988. Case-controlled study of childhood cancer and exposure to 60-Hz magnetic fields. Am. J. Epidemiol. 128:21–38.

Schafer, C. R. 1977. Hearing aid. U.S. patent no. 4,052,572 (4 Oct.).

———. 1980. Hearing aid with modulated suppressed carrier signal. U.S. patent no. 4,220,830 (2 Sept.).

———. 1987. Cortical hearing aid. U.S. patent no. 4,711,243 (8 Dec.).

Schelkunoff, S. A. 1943. Electromagnetic waves. New York: D. Van Nostrand.

Schiff, L. I. 1949. Quantum mechanics. New York: McGraw-Hill.

Schnorr, T.; Grajewski, B.; Hornung, R.; Thun, M. J.; Egeland, G.; Murray, W.; Conover, D.; Halperin, W. 1991. Video display terminals and the risk of spontaneous abortion. New Eng. J. Med. 324:727–33.

Schwan, H. P. 1983. Biophysics of the interaction of electromagnetic energy with cells and membranes. In Grandolfo, M.; Michaelson, S. M.; and Rindl, A., eds. Biological effects and dosimetry of nonionizing radiation. New York: Plenum, 213–31.

———. 1988. Biological effects of non-ionizing radiations: Cellular properties and interactions. Annals Biomed. Eng. 16:245–78.

Scott, A. C. 1982. Dynamics of Davydov solitons. Phys. Rev. A26:578.

Shedd, T. C. 1974. Encyclopaedia Britannica, 15th ed., s.v. "Railroads and locomotion."

Smith, S. D.; McLeod, B. R.; Liboff, A. R.; Cooksey, K. E. 1987. Calcium cyclotron resonance and diatom motility. Bioelectromagnetics 8:215–27.

Smyth, H. de W.; Ufford, C. W. 1939. Matter, motion and electricity. New York: McGraw-Hill.

Smythe, W. R. 1950. Static and dynamic electricity. New York: McGraw-Hill.

Snell, R. S. 1978. Atlas of clinical anatomy. Boston: Little, Brown.

Spiegel, R. J. 1977. High-voltage electric coupling to humans using moment method techniques. IEEE Trans. Biomed. Eng. BME-24:466–72.

Spiegel, R. J.; Joines, W. T.; Blackman, C. F. 1982. Calcium-induced efflux from isolated brain tissue: Is it caused by an electromagnetically induced pressure wave? Abs. fourth annual mtg. Bioelectromagnetics Soc.

Stein, G. S. 1992. Effects of electric and magnetic fields on growth control. Oak Ridge, Tenn.: ORAU, IV-1–IV-24.

Stratton, J. A. 1941. Electromagnetic theory. New York: McGraw-Hill.

Stuchly, M. A.; Ruddick, J.; Villeneuve, D.; Robinson, K.; Reed, B.; Lecuyer, D. W.; Tan, K.; Wong, J. 1988. Teratological assessment of exposure to time-varying magnetic field. Teratology 38(4):461–66.

Sunde, E. D. 1968. Earth conduction effects in transmission systems. New York: Dover.

Tabrah, F. L.; Guernsey, D. L.; Chou, S.-C.; Batkin, S. 1978. Effect of alternating magnetic fields (60–100 gauss, 60 Hz) on *Tetrahymena pyriformis*. T.I.T. J. Life Sci. 8(3–4):73–77.

Tamarkin, L.; et al. 1982. Decreased nocturnal plasma melatonin peak in patients with estrogen receptor positive breast cancer. Science 216:1004–05.

Tell, R. A.; et al. 1977. Examination of electric fields under EHV overhead power transmission lines. Washington, D.C.: U.S. Environmental Protection Agency, Office of Radiation Programs. EPA-520/2-76-008.

Tolman, R. C. 1946. The principles of statistical mechanics. Oxford: Oxford University Press.

Tomenius, L. 1986. 50-Hz electromagnetic environment and the incidence of childhood tumors in Stockholm county. Bioelectromagnetics 7:191.

Trichopoulos, D. 1992. Epidemiological studies of cancer and extremely low-frequency electric and magnetic field exposures. ORAU, Oak Ridge, Tennessee, pp. v1–v58.

Verveen, A. A.; Derksen, H. E. 1968. Fluctuation phenomena in nerve membranes. Proc. IEEE 56:906.

Watson, J. 1979. The electrical stimulation of bone healing. Proc. IEEE 67:1339–51.

Weaver, J. C. 1992. Electromagnetic field dosimetry: Issues relating to background noise and interaction mechanisms. Bioelectromag. Supp. 1:115–17.

Weaver, J. C.; Astumian, R. D. 1990. The response of cells to very weak electric fields: The thermal noise limit. Science 247:459–62.

———. 1992. Estimates for ELF effects: Noise-based thresholds and the number of experimental conditions required for empirical searches. Bioelectromag. Supp. 1:119–38.

Wertheimer, N.; Leeper, E. 1979. Electrical wiring configurations and childhood cancer. Am. J. Epidem. 109:273–84.

Wilson, B. W.; Anderson, L. E.; Hilton, D. I.; Phillips, R. D. 1981. Chronic exposure to 60-Hz electric fields: Effects on pineal function in the rat. Bioelectromagnetics 2:371–80.

Wilson, B. W.; Chess, E. K.; Anderson, L. E. 1986. Sixty-Hz electric-field effects on pineal melatonin rhythms: Time course for onset and recovery. Bioelectromagnetics 7:239–42.

Index